SMALL-BLOCK CHEVY ENGINE BUILDUPS

How to Build Horsepower for Maximum Street and Racing Performance

Covers All Makes and Models

From the Editors of *Chevy High Performance* Magazine

HPBOOKS

HPBOOKS
Published by the Penguin Group
Penguin Group (USA) Inc.
375 Hudson Street, New York, New York 10014, USA
Penguin Group (Canada), 90 Eglinton Avenue East, Suite 700, Toronto, Ontario M4P 2Y3, Canada
(a division of Pearson Penguin Canada Inc.)
Penguin Books Ltd., 80 Strand, London WC2R 0RL, England
Penguin Group Ireland, 25 St. Stephen's Green, Dublin 2, Ireland (a division of Penguin Books Ltd.)
Penguin Group (Australia), 250 Camberwell Road, Camberwell, Victoria 3124, Australia
(a division of Pearson Australia Group Pty. Ltd.)
Penguin Books India Pvt. Ltd., 11 Community Centre, Panchsheel Park, New Delhi—110 017, India
Penguin Group (NZ), 67 Apollo Drive, Rosedale, North Shore 0745, Auckland, New Zealand
(a division of Pearson New Zealand Ltd.)
Penguin Books (South Africa) (Pty.) Ltd., 24 Sturdee Avenue, Rosebank, Johannesburg 2196, South Africa

Penguin Books Ltd., Registered Offices: 80 Strand, London WC2R 0RL, England

While the author has made every effort to provide accurate telephone numbers and Internet addresses at the time of publication, neither the publisher nor the author assumes any responsibility for errors, or for changes that occur after publication. Further, publisher does not have any control over and does not assume any responsibility for author or third-party websites or their content.

SMALL-BLOCK CHEVY ENGINE BUILDUPS

Copyright © 2003 by Primedia, Inc.
Book design and production by Michael Lutfy
Cover design by Bird Studios
Cover photos by *Chevy High Performance* magazine
Interior photos by author unless otherwise noted

All rights reserved.
No part of this book may be reproduced, scanned, or distributed in any printed or electronic form without permission. Please do not participate in or encourage piracy of copyrighted materials in violation of the author's rights. Purchase only authorized editions.
HPBooks is a trademark of Penguin Group (USA) Inc.

First edition: January 2003

ISBN: 978-1-55788-400-8

PRINTED IN THE UNITED STATES OF AMERICA

20 19 18 17 16 15 14

NOTICE: The information in this book is true and complete to the best of our knowledge. All recommendations on parts and procedures are made without any guarantees on the part of the author or the publisher. Tampering with, altering, modifying or removing any emissions-control device is a violation of federal law. Author and publisher disclaim all liability incurred in connection with the use of this information. The information contained herein was originally published in *Chevy High Performance* magazine and is reprinted under license with Primedia, Inc., 6420 Wilshire Blvd., Los Angeles, CA 90048. *Chevy High Performance* is a trademark of Primedia Specialty Group, Inc. and is used with permission. Copyright © 2003 Primedia Specialty Group, Inc. All rights reserved.

CONTENTS

Introduction		v
Chapter 1:	How to Build Your First Engine	1
Chapter 2:	The Goodwrench Quest, Part 1	6
Chapter 3:	The Goodwrench Quest, Part 2	12
Chapter 4:	The Goodwrench Quest, Part 3	17
Chapter 5:	The Goodwrench Quest, Part 4	23
Chapter 6:	The Goodwrench Quest, Part 5	29
Chapter 7:	The Goodwrench Quest, Part 6	35
Chapter 8:	The Goodwrench Quest, Part 7	41
Chapter 9:	Street Fighter 377	46
Chapter 10:	Agent 87, Part 1	51
Chapter 11:	Agent 87, Part 2	56
Chapter 12:	Agent 87, Part 3	61
Chapter 13:	Hot, Hot, Hot Cam	68
Chapter 14:	Son of Muscle Mouse	76
Chapter 15:	Gladiator vs. Muscle Mouse	81
Chapter 16:	Battle of the Small-Block Strokers	86
Chapter 17:	Flow to Go	99
Chapter 18:	Flow Power	117
Chapter 19:	Bolting On Vortec Heads	121
Chapter 20:	A Tale of Torque	125
Chapter 21:	Porting For Power	129
Chapter 22:	Compression Lessons	134
Chapter 23:	New Wave TPI	139
Chapter 24:	It's a Spring Thing	143
Chapter 25:	Angling for Power	147
Chapter 26:	Cam Basics	151
Chapter 27:	Cam Overlap	157
Chapter 28:	Roller Cam Basics	162
Source Index		166

ACKNOWLEDGMENTS

Although this publication is a team effort between Primedia and HPBooks, it is made possible by the individual efforts of Craig Nickerson, President of Primedia, Inc.; Dave Cohen, Vice President of Sales and Marketing; Jeff Smith, Editor of *Chevy High Performance* magazine; Sean Holzman, Martha Guillen and Eric Goldman, Primedia Licensing. The following writers, editors and photographers also need to be acknowledged for their contributions that follow: Jeff Smith, Ed Taylor, Mike Petralia, and John Baechtel.

INTRODUCTION

ON SMALL-BLOCKS AND CAR GUY CHEMISTRY

There's some kind of special chemistry between a car guy and an engine. Make that motor a small-block Chevy with an aluminum intake, four-barrel carburetor, and a set of headers, and you've just created the automotive equivalent of Love Potion Number Nine. It must be the 30-weight engine oil that gets into the pores of your skin that causes guys to get all excited about a small-block Chevy. Whatever it is, car guys love nothing better than messin' with motors. When it came baptism time, you were probably like us and asked the motor minister to dunk you long and deep in those horsepower waters. Yeah, baby. It's all about horsepower.

Okay, so we've established that horsepower is our weakness. We admit it. We have to resist the urge to act like a 10 year-old and go charging out the door when a heavily cammed small-block rumbles down the street. And we've been known to skip both lunch and dinner so we can spend all day in the garage fiddling with bearing clearances until we've got exactly 0.0024 to 0.0026-inch of clearance for all eight rods. We're not talking about a race motor here, that's just the effort we put into the daily-driver engine for the wife's grocery getter. The thumper motors deserve even more time.

You probably have your own stories to tell about your favorite small-block. The amazing thing is that everybody's story is the same and yet completely different. We all have friends who to this day contend that the best engine Chevrolet ever built was the 327. Others believe that the LS1 will create the next new dynasty. And they may be right. Still others will go to their grave contending that the 400 was the best small-block because it was the biggest. They also believe the 400 should have been even bigger and revere those hardy few who have built 454ci small-blocks.

Frankly, we love 'em all. We've built 283's and 420ci small-blocks. The big-inch motors are fun because they sound like a Rat and it's fun to see the look on a guy's face when you pop the hood and he stares at a small-block where he was sure he'd see a 468-inch Rat. It's even more fun when your finely-tuned small-block squeaks out a win over a healthy Rat and big-block boy has to try and explain how some little Mouse motor beat up on his fat-block. The other side of the coin is taking a mini small-block like a 283 or a 307 and turn it into a nice performing street engine that can knock down 22 mpg and still run 14's in the quarter-mile. Trust us, it can be done.

The point to all this is there are a million ways to build a small-block. This book has merely scratched the surface. Since we completed this compilation, we've learned even more and a dozen or so new pieces have jumped into the small-block Chevy parts pool to make even more power. Better yet, the future looks downright powerful when you consider combining a small-inch 302ci Mouse motor with a powerful EFI system and a pair of turbos. We're talking about an engine that could knock down over 20 mpg and make an easy 700 hp. Yes sir, we'll be making serious horsepower in the near future with a small-block that could live for thousands of miles of street driving. And it's all because that car-guy chemistry just won't leave it alone. How bad do you have it?

Jeff Smith
Editor
Chevy High Performance

How To Build Your First Engine

Doing it Right the First Time
by Jeff Smith
Photography by Jeff Smith and Ed Taylor

Engines and horsepower are what hot rodders live for, and nothing's more gratifying than building the engine yourself. There's nothing more satisfying than hearing that engine run for the first time after you've spent hours carefully massaging the powerplant. If you've never built an engine before, we're here to help you get there. It's really not that difficult. It all comes down to spending a little more time to line up all the tools you'll need along with the patience to assemble it correctly.

Chances are you won't have all the tools necessary to perform all the recommended operations in this story. That doesn't give you a back door to skip these procedures. That just means you'll need to beg, borrow, or rent these tools and learn how to use them properly. If you have buddies who are engine builders, then they probably have the tools and would be willing to show you how to use them. Maybe you could also talk them into looking over your shoulder as you assemble your engine. It's worth the effort, and you can always trade for their help by offering to help work on one of their projects.

We'd also suggest reading other stories on engine building, and buying books on building a small-block Chevy. The books can go into much more detail than we can in this story, and it's the details that will make your engine buildup completely successful.

We've covered the major points here that, if overlooked, can cause engine failure. We did not go into detail on starting the engine for the first time, but this is a critical part as well.

For starters, always pre-lube the engine. You can build a pre-luber out of an old distributor body and shaft. Also make sure the ignition timing is set at about 12-degrees BTDC and the carb is primed. That way the engine should only have to turn over once or twice before it fires. Then immediately bring the engine up to 2,000 to 2,500 rpm, varying the rpm for the first 15 minutes. Monitor the coolant temperature and oil pressure and you're on your way.

So cruise through this how-to-build-it story and start planning your next engine. Then you can proudly claim, "Yeah, I built this myself" when the cruisin' spectators ask, "Who built the engine?" It's a great feeling of accomplishment.

Small-Block Chevy Engine Buildups

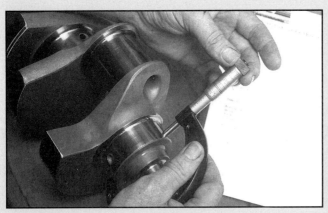

The only way to know the bearing clearance for the mains and rods is to measure the outside diameter of the main and rod journals with a micrometer and write them down. Then measure the inside diameter of the main bearing and connecting-rod housing bores with the bearings in place. If the bearing clearance is either too tight or too loose, you will need to make adjustments. Do not rely on Plastigage to set your bearing clearance. It's not accurate enough.

The best way to ensure proper clearances is to pre-assemble the rotating assembly. For example, checking the crankshaft endplay can be done with new bearings after the crank is turned. If the clearance is too tight, you can massage the bearings to add clearance. A trick is to torque all the main caps in place but leave the thrust (No. 5) cinched but not tight. Then tap the crank forward to align the thrust washer. Then torque the No. 5 main.

Pre-assembling also gives you a chance to check rod-side clearance. If you find a pair of rods that are too tight, you can swap cylinders to gain clearance. This is acceptable as long as you swap rods with cylinders on the same side of the engine so that the chamfer on the rod faces outboard on the crank.

You can also check deck height during pre-assembly. If the piston is too far down in the hole, you still have the opportunity to have the block decked.

You can expect to spend at least an hour cleaning the block. Do this right before assembly.

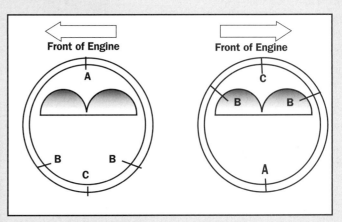

Once the cam and crank are in place and the mains torqued, you can assemble the rings on the pistons based on these drawings. The idea is to space the ring end gaps so they don't line up and create a leakage path. **A**—Oil ring expander and top-ring gaps. **B**—Oil ring scraper-rail gaps. **C**—Second-ring gap

How To Build Your First Engine

Some builders like ring-spreader tools, but we've broken rings with these. We prefer to spiral the rings on the pistons. Install the oil rings first, then the second ring, followed by the top compression ring.

The best tool for installing pistons is the tapered ring compressor. Powerhouse, ARP, and others sell the tools that make installing pistons a breeze.

These rod-bolt protectors from ARP align the rod over the crank. Those small plastic covers also work well to prevent nicking the crank.

The only way to properly torque the rod bolts is with a rod-bolt-stretch gauge. Typical ARP bolt stretch is around 0.006 inch.

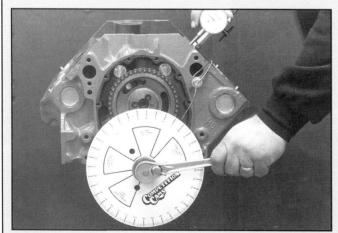

With the timing chain and gears installed, this is a great time to degree the cam before the heads are installed. With today's high-quality parts, we've found that the majority of the time the cam is right where it's supposed to be.

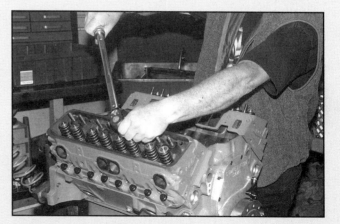

Torque the heads on in a circular fashion starting with the center head bolt. Most engine builders prefer to torque the head bolts in three stages, starting at 35 ft-lbs, then 45 ft-lbs, and finally 60 ft-lbs of torque. Be sure to coat the bolt threads with sealant to prevent leaks.

Small-Block Chevy Engine Buildups

If you don't have a special tool to install the harmonic damper, you can use a longer fine-thread crank bolt to pull the damper in place. You could also use a large hammer and a block of wood, but that's troglodyte tech. You know better than that. This tool comes from Powerhouse.

Be sure to seal the intake gaskets to the heads using Permatex sealant. Then you can seal the intake end rails using a bead of Permatex silicone rather than the supplied end-rail gaskets that tend to push out and leak.

After the intake is installed, always pre-lube the engine before firing it up. This is great insurance and worth the effort.

CLEARANCES

A good engine builder is someone who checks all the critical clearances to ensure that the engine is assembled properly. The following are suggestions for proper clearances on a 4-inch bore small-block Chevy. Piston-to-wall clearance is always specified by the piston manufacturer. The larger clearance for the No. 5 main bearing allows more oil to pass through the bearing to lubricate the thrust bearing.

Component	Clearance (inches)
Main bearing	0.002–0.0025 mains 1-4
	0.0025–0.0030 main 5
Rod bearing	0.002–0.0025
Rod side	0.009–0.013
Piston to wall	Refer to piston mfr.
Crank endplay	0.003-0.010
Ring end gap	Refer to piston mfr.
	0.016 top
	0.022 second
Piston to head	0.040 (steel rod)
Valve to piston	0.080 intake
	0.100 exhaust
Valve to guide	0.0015 intake
	0.0015–0.0020 exhaust
Retainer to seal	0.050
Coil bind	0.050

How To Build Your First Engine

POWER STEPS

The following is an abbreviated step-by-step outline for final assembly of a small-block Chevy. This is a generic procedure, but it will help you in assembling your first engine.

1. Clean the block and install all pipe and freeze plugs.
2. Clean the cylinder walls with paper towels and ATF.
3. Install the camshaft using moly paste on a flat-tappet cam.
4. Insert the main bearings and lube.
5. Drop in the crank and cinch the main caps.
6. Tap the crank forward with a rubber mallet to set thrust bearing. Torque the main caps.
7. Install rings on the pistons, position the ring end gaps, and install rod bearings.
8. Cover rod bolts and use a tapered-ring compressor to install the piston.
9. Lube rod bolts and use a stretch gauge to torque the rod bolts.
10. Install the oil-pump driveshaft and torque the oil-pump stud or bolt.
11. Line up the marks on the timing gears and install the cam drive assembly.
12. Degree the camshaft. Torque the cam bolts using a thread locker.
13. Install the timing chain gasket, cover, and ignition-timing indicator.
14. Install the oil-pan gasket and oil pan.
15. Turn the engine right-side up and install the head gaskets and heads.
16. Put sealant on the head bolts and torque in three steps—35–45–60 ft-lbs.
17. Lube lifter faces and install along with the pushrods.
18. Lube rocker balls with moly and set them in place.
19. Set valve lash with a half-turn preload (hydraulic lifters).
20. Place RTV on the end rails of the block and glue the intake gasket firmly to each head.
21. Place intake manifold on the engine and start all bolts. Install distributor and tighten all bolts.
22. Fill the oil filter and engine with oil.
23. Remove the distributor and install pre-oiler to pressure-lube the engine.
24. Connect the oil-pressure gauge and pressure-lube the engine.
25. Replace the distributor and static-time the engine using 12 degrees BTDC.
26. Install the carburetor, prime it with fuel, and hook up the exhaust system.
27. Fire the engine and maintain at least 2,000 to 2,500 rpm for the first 15 minutes of engine run time. Watch the coolant temperature and oil pressure closely for signs of problems.
28. Test-drive your new engine and enjoy the ride!

TORQUE SPECS

Component	Torque (ft-lbs)
Main cap	Two-bolt: 70
	Four-bolt Inner: 70
	Outer: 65
Connecting-rod bolt	45–50 (stock 3/8 inch)
Head bolts	65
Rocker stud	50
Oil-pump stud (ARP)	45 w/moly lube
	50 w/oil
Cam sprocket	20
Intake manifold	30
Flywheel	60

2 The Goodwrench Quest, Part 1

By Jeff Smith
Photography by Ed Taylor

In the world of small-block Chevys, there are literally hundreds of engines to choose from. The choices range from cheapie backyard rebuilds to $10,000-plus race motors. Among all of these engines clamoring for your attention, there is one that may be the deal of the century. For years, Chevrolet has offered the Goodwrench 350 as a brand new service replacement engine. While you can purchase this engine from any GM dealer in the country, there are a few dealers like Scoggin-Dickey that are offering Mr. Goodwrench for the outstanding price of $1,190! This isn't some slapped-together rebuild, it's an all new 350 long-block assembled with new parts. And better still, it comes with a warranty.

If you're looking for a basic small-block that will deliver years of trouble-free operation, you'll be hard-pressed to find a better buy. Because we live and breathe Chevy high performance, we thought it would be fun to see what kind of voodoo we could do to extract more grunt from our Goodwrench Mouse. The plan was simple: borrow a Goodwrench 350 engine from Scoggin-Dickey, bolt it on the dyno, and run the wee out of it.

This first of several episodes started with a test of the basic engine package with a factory Q-jet intake, cast-iron exhaust manifolds, HEI ignition, and a full dual-exhaust system. The next test added a set of Hooker headers and finished with the addition of an

The Goodwrench Quest, Part 1

The Goodwrench 350 engine comes as a long-block assembly complete with an oil pan, timing chain cover, and valve covers. This Goodwrench motor came from Scoggin-Dickey, delivered directly to our doorstep. Complete with a three-year/50,000-mile warranty, the price was right at $1,190 plus shipping.

This is a collection of the small parts we needed to complete the Goodwrench and get it running. A water pump, a fuel pump, a Fel-Pro intake gasket, a fuel pump mounting plate, a distributor hold-down, an ARP balancer bolt and fuel pump pushrod, and a timing indicator all came out of the PAW catalog. The oil filter adapter, dipstick, and tube all came from our local Chevy dealer. This particular HEI is a bone-stock distributor from GM Performance Parts. The only other miscellaneous parts you'll need will be oil, a filter, spark plugs, plug wires, and fuel.

Whenever you start a new engine for the first time, it's best to pressure-lube it. We used a 1/2-inch drill motor, a homemade pre-luber, and an oil pressure gauge. With the oil and filter in place, spin the oil pump clockwise until you read oil pressure on the gauge. It's also a good idea to bump the engine over a couple of times while doing this to ensure everything is well-lubed.

In addition to the cast-iron 2-inch exhaust manifolds, our test included the right side heat riser valve as well. This is necessary to connect the stock exhaust pipes, but you could drill out the valve itself to improve flow. We chose to leave it in place for our test.

Edelbrock Performer intake. All the tests were run on 92-octane pump gas.

THE ENGINE

The Goodwrench 350 is unquestionably bread-and-butter basic. While all production small-blocks have employed a one-piece rear-main seal design since 1986, the Goodwrench 350 is assembled in Mexico and retains the classic two-piece pre-'86 design. This makes it an excellent choice for a basic hot rod motor for all pre-'86 cars since you don't have to purchase a new flexplate or flywheel.

Small-Block Chevy Engine Buildups

PARTS LIST

These are the parts we used to complete the Goodwrench engine for the testing.

Component	Source	Part Number
Goodwrench 350	Scoggin-Dickey	10067353
HEI distributor	GMPP	1104067
Flexplate	GMPP	471598
Harmonic balancer	GMPP	12555879
Timing tab	GMPP	12341915
Oil filter adapter	GMPP	3952301
Intake manifold gasket	Fel-Pro	1256
Water pump	PAW	TRW-FP-1521
Fuel pump	Carter	M6900
Fuel pump mounting plate	PAW	WYSWA-2310
Harmonic balancer bolt	ARP	134-2501
Fuel pump pushrod	ARP	134-8701
Distributor hold-down	PAW	CHR-7606
Dipstick and tube	PAW	CHR-7171
Spark plug wires	MSD	3121
Spark plugs	AC	R-44TS
Headers, '68 Nova	Hooker	2451
Mufflers, 2 1/4-inch	Hooker	21005

The Hooker headers are standard chassis headers for early Novas but represent a typical street header with 1 5/8-inch primary pipes and a 3-inch collector. We also intend to test the intermediate-length headers for this engine.

The baseline induction system consisted of a factory aluminum intake and a dependable Q-jet. The intake is an LG-4 305 style that has a decent reputation, but the only place it did well was under 2,700 rpm. This time-honored Q-jet has proven itself over many years of trouble-free operation. What many Chevy enthusiasts don't know is that the Q-jet is rated at 750 cfm. At the 260 hp level, however, the engine is barely using more than about 500 cfm.

Ignition duties were handled by an out-of-the-box Chevy HEI distributor along with a set of MSD wires and Bosch spark plugs.

Starting with a four-bolt main cap block, the 350 employs a standard cast crank and cast-aluminum flat-top pistons with ductile-iron 5/64-inch rings. Chevy claims the compression is a wheezy 8.1:1, but after the test was over, we measured everything and came up with a slightly better 8.4:1. The good news with this low compression is that the Goodwrench engine should even run on 87-octane gas. The downside is that this low compression certainly sacrifices power. The long-block comes complete with an oil pump and pan, as well as a timing-chain cover and valve covers. The cam is a simple flat-tappet hydraulic with specs that are bone-stock tame. The 76cc chamber cast-iron heads (casting number 83417368) are fitted with 1.94/1.50-inch intake and exhaust valves and stock stamped-steel rockers.

Beyond the mechanical aspects, there's also the GM warranty. According to Scoggin-Dickey, GM offers a three-year, 50,000-mile warranty on the engine as long as it has not been internally altered; this means that the warranty would be void if a performance camshaft was added or if the compression was increased. However, a performance intake manifold and/or headers are acceptable modifications. There are other warranty details too numerous to mention here, which you should investigate if you are considering purchasing one of these engines.

The Goodwrench Quest, Part 1

We used a Fel-Pro 1256 intake manifold gasket to seal each intake to the engine. The silicon rings around the intake and water passage ports seal the ports to prevent vacuum leaks.

The Edelbrock Performer intake made a world of difference in our Goodwrench 350. Combined with the Hooker headers, we saw an excellent 265 hp at only 4,300 rpm and 350 lb-ft of torque at 3,600. The combination of the intake and headers was worth as much as 59 lb-ft of torque at 3,400 rpm!

All dyno work was performed at Ken Duttweiler Performance in Saticoy, California. All tests were performed using 92-octane pump gas, but at 8.1:1, we probably would have seen similar (if not better) results with 87-octane fuel.

With shipping costs ranging from $125 to $165, depending on how far the engine must travel, you could have a Goodwrench 350 at your doorstep for $1,355 or less. One other important point is that Scoggin-Dickey buyers outside the state of Texas do not have to pay sales tax, which can represent a savings in excess of $100.

DYNO TESTING

Freelancer, engine builder, and CHP's man-about-town, Ed Taylor, ram-rodded the Goodwrench Mouse project for us at Ken Duttweiler Performance in Saticoy, California. The first test was to cork the 350 with a stock aluminum intake manifold, a Q-jet carburetor, and cast-iron exhaust manifolds connected to a pair of Hooker 2 1/4-inch turbo-style mufflers. As you can see from the Test 1 results, the engine performed much better than GM's stock 190 hp rating, making 239 hp at 4,300 with peak torque coming in at 3,700 rpm with 324 lb-ft. This is a rather narrow powerband between peak torque and peak horsepower, and it was obvious that both the intake and exhaust systems were extremely restrictive. Remember, this is a stone-stock 350 with kerosene-compatible compression and a camshaft that barely bumps the valves off their seats. On the plus side, the engine made more than 250 lb-ft of torque from 2,500 to 4,800.

It was obvious the engine needed to breathe, so we pitched the cast-iron exhaust manifolds and bolted on a set of Hooker 1 5/8-inch headers. We retained the 2 1/4-inch exhaust pipes

Cast-iron exhaust manifolds are incredibly restrictive, even on a stock engine like the Goodwrench 350. The iron manifolds are connected with a set of 2 1/4-inch dual pipes bolted to a set of Hooker mufflers.

and Hooker turbo mufflers, all obtained from PAW. The engine responded immediately to the exhaust system upgrade with as much as 53 lbs-ft more torque at 3,400 rpm! To put that in perspective, that's like a mild shot of nitrous. As you might expect, horsepower also improved, with 17 more horsepower at 4,500 rpm. We were certainly on the right track, but now the engine needed help with the induction side.

The route to adding more horsepower was obvious—the Goodwrench motor needed a better intake manifold. So, we unbolted the factory aluminum piece and bolted on an Edelbrock Performer intake. The Performer is drilled for both the square-flange Holley-style bolt pattern and the spread-bore Q-jet pattern, which makes installation easy. Within a few minutes, Taylor fired the engine back up and ran it through the 2,500- to 5,300-rpm power test. Looking at the power curves, the Performer lost a little power below 3,000 rpm, but more than made up for it with improved power in the midrange. This is probably not a problem with the manifold as much as it reflects a calibration difficulty with the Q-jet that could have been solved with a bit more tuning. But overall, the combination of the Goodwrench 350, a set of 1 5/8 headers, an Edelbrock Performer intake, and a stock Q-jet was worth 350 lb-ft of torque and 265 hp. Not bad for a stock motor and bolt-on parts!

WHERE DO WE GO FROM HERE?

In the next chapter, we will pocket-port the stock heads and then try a better camshaft. In the chapters that follow, we've also got a set of aluminum Corvette heads and a set of Vortec production iron heads we're going to swap into place as well as other parts combinations, bigger cams, different carburetors, and even some variations on the header theme. The Vortec heads and other GM hi-po parts come from GM Performance Parts, or can also be obtained directly from Scoggin-Dickey. So stick with us—we plan on beating the snot out of this Goodwrench crate engine to find out just how much power we can squeeze out of a budget small-block.

JUST TESTING

The following chart illustrates the power of the Goodwrench 350 in its three basic configurations. Note that all differences indicated in this chart are in comparison to Test 1. Test 3 looks better in the difference column because it represents the combination of both the headers and the intake manifold.

TEST 1: Stock Goodwrench 350 with stock Q-jet aluminum intake, Q-jet carburetor, cast-iron exhaust manifolds and 2 1/4-inch dual-exhaust with Hooker turbo-style mufflers. Ignition timing was set at 34 degrees with 92-octane pump gas.

TEST 2: Same as above except Hooker 1 5/8-inch long-style primary pipe street headers replaced the iron exhaust. Ignition timing remained the same.

TEST 3: Same as above except we added an Edelbrock Performer intake manifold. Timing and jetting remained the same.

RPM	TEST 1 TQ	TEST 1 HP	TEST 2 TQ	TEST 2 HP	DIFF TQ	DIFF HP	TEST 3 TQ	TEST 3 HP	DIFF TQ	DIFF HP
2500	278	132	267	127	-11	-5	255	121	-23	-11
2600	272	135	283	140	11	5	267	132	-5	-3
2700	287	147	289	148	2	1	265	136	-32	-11
2800	288	154	299	159	11	5	274	146	-14	-8
2900	288	159	309	170	11	11	274	151	-14	-8
3000	283	162	303	173	20	11	274	157	-9	-5
3100	276	163	300	177	24	14	277	163	1	0
3200	266	162	292	178	26	16	277	168	11	6
3300	261	164	294	185	33	21	291	183	30	19
3400	271	176	324	210	53	34	330	214	59	38
3500	289	193	306	204	17	11	347	231	58	38
3600	320	219	349*	239	17	20	350*	240	30	21
3700	324*	228	342	241	18	13	350	246	26	18
3800	321	232	338	245	17	13	347	251	26	19
3900	316	235	336	249	20	14	344	256	28	21
4000	311	237	330	251	19	14	338	257	27	20
4100	306	239*	324	253	18	14	333	260	27	21
4200	298	239	317	253	19	14	329	263	31	24
4300	292	239	311	255	19	16	323	265*	31	26
4400	285	238	304	255*	19	17	315	264	30	26
4500	277	237	296	254	19	17	306	262	29	25
4600	270	237	286	250	16	13	296	259	26	22
4700	261	233	277	248	16	15	288	258	27	25
4800	253	231	267	244	14	13	277	253	24	22
4900	244	228	259	241	15	13	267	250	23	22
5000	238	227	248	236	10	9	259	247	21	20
5100	229	222	239	232	10	10	250	242	21	20
5200	219	217	229	227	10	10	240	237	21	20
5300	210	212	218	220	8	8	231	233	21	21

*Denotes peak

3 The Goodwrench Quest, Part 2
By Jeff Smith
Photography by Ed Taylor

Budget horsepower is the name of this game. If you read the exploits of our GM Goodwrench budget 350 provided by Scoggin-Dickey Performance Center in the last chapter, then you know that we pumped a few steroids into the stock OEM combination in the form of a set of headers and an Edelbrock Performer intake. In this configuration through mufflers, we made a respectable 265 hp and an even stronger 350 lbs-ft of torque. While that's not bad for a couple of simple bolt-ons, we're not nearly satisfied. We know this pedestrian 350 is ca-pable of much more power.

The simplest path to increased power is to methodically remove the restrictions that prevent the engine from breathing more deeply. We started that process in the last chapter with the headers and intake. Now the cork in the system was the cylinder heads. While we could just bolt on a set of better heads (which we will do later), we thought we'd first try improving the existing cylinder heads. This was also the cheapest way to go.

Our first step was to remove the cylinder heads and do a little measuring. Chevy claims the engine squeezes a mean 8.1:1 compression, but somehow, in a fog of mathematical befuddlement, we claimed in the last chapter that the Goodwrench motor measured 8.4:1 compression. We were wrong. After we rechecked our ciphering, Mr. Goodwrench proved to be much lazier. The actual compression figures out to be a menial 7.8:1! Based on this complacent squeeze factor, we decided that increasing compression and improving airflow would be a good idea.

FLOW TO GO

As you might imagine, stock 1.94/1.50-inch castings are weak when it comes to air flow. In the name of budgetary restraint, we decided to stick with the stock valves so we could evaluate the effect of the porting. McKenzie Cylinder Heads in Oxnard, California, performed the cylinder head porting for this story, but per our instructions, Todd McKenzie spent very little time on these heads. Better yet, you can duplicate this work yourself with little more than a porting kit from Standard Abrasives, a small air compressor, and a hardware store 1/4-inch die grinder. The plan was to perform a simple pocket port job that would generate the most results for the least effort.

If this is your first attempt at cylinder head porting, you may want to practice first on a junk cylinder head. While the work looks intimidating, it really is easy to do. In fact, once you get comfortable, the hardest part is to stop working after blending the seat area 1 inch past the bottom of the valve seat. The key is to stay conservative and not hog out the port in search of more power.

Another area worth investigating is back-cutting the valves. Flow bench testing has proved that placing a 30-degree back cut on both intake and exhaust valves improves the low- and mid-lift flow of both the intake and exhaust ports. This back cut is placed just inside the 45-degree face of the valve and acts much like the three-angle radius of a valve seat to improve flow. The 30-degree back cut can be accomplished by any competent machine shop and will improve

PARTS LIST

Component	Source	Part Number
Goodwrench 350	Scoggin-Dickey	10067353
Performer intake	Edelbrock	2101
Headers, 1 5/8-inch	Hooker	Application specific
Mufflers	Hooker	21005
Camshaft	Comp Cams	XE268H-10
Cam and kit	Comp Cams	K12-242-2
Head bolts	ARP	134-3601

Fel-Pro makes a great rubber-embossed steel shim gasket that is only 0.015-inch thick, which increases the compression on engines like this one. This is a great gasket that seals well and also improves power by reducing the quench area above the piston.

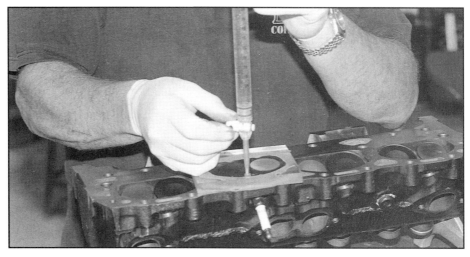
We cc'd the heads before and after to determine the actual combustion chamber volume to work out the static compression ratio. Milling the heads only reduced the chamber volume by 3 cc.

Small-Block Chevy Engine Buildups

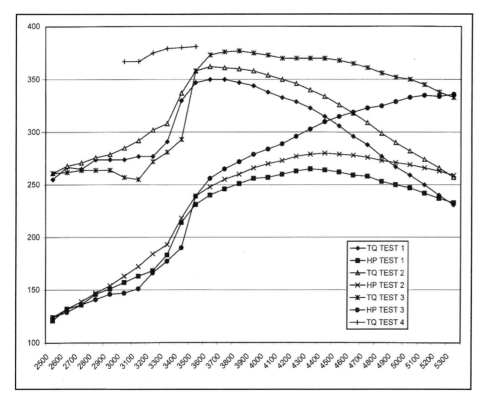

As the graph reveals, the combination of a pocket-ported head and the camshaft increased torque and horsepower dramatically. The short torque curve in the upper left indicates what the low speed torque should have been had we repaired the Q-jet's fuel curve problem. The dip in the torque curve is purely a fuel curve situation.

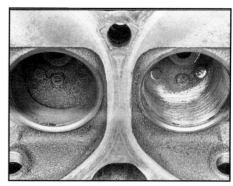

Our quest for more power started with the cylinder heads. Todd McKenzie did some minimal pocket-port work (right side) on the stock iron castings and improved the port flow, especially on the exhaust side. This is easy work that you can do at home in roughly eight hours. We retained the stock 1.94/1.50-inch valves just to see how much we could improve without spending money on bigger valves. A stock intake port is on the left.

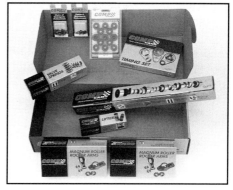

The hot ticket is to purchase the Comp Cams K-kit when you buy the cam. The kit comes with the cam, lifters, retainers, keepers, springs, Magnum rocker arms, and a timing set.

If you choose to do your own home porting, the best place to start is with a Standard Abrasives Deluxe Engine Porting Kit. This provides everything you need, except for the die grinder. If you want to speed the process, a carbide cutter would also be a good investment.

McKenzie did back cut the exhaust valves with a 30-degree angle (right) just inside the 45-degree seat. This dramatically improves port flow at low lift numbers, helping scavenge the cylinder when the valve first opens.

cylinder head airflow even if pocket porting is not included. We have found that back cutting both the intake and exhaust works, but McKenzie decided to back cut only the exhaust side for this test.

Before the heads were reassembled, McKenzie also treated them to a simple milling operation that reduced the chamber volume from 76 cc to 73 cc. The effort was aimed more at ensuring that the deck surface was flat than at attempting to dramatically increase compression. However, we also decided to eliminate the stock GM composition gasket that measured a stout 0.039-inch thick in favor of a much thinner 0.015-inch-thick gasket from Fel-Pro.

After measuring everything again and doing accurate ciphering, we came up with a slightly better 8.4:1 compression. While this is hardly killer squeeze, we figured every little bit helps.

The Goodwrench Quest, Part 2

The factory spring tested at 75 pounds of tension at the installed height of 1.70 inches. The Comp Cams spring tested at a much better 110 pounds on the seat at the same installed height.

CAM SPECS

This chart compares the stock GM Goodwrench cam specs to the Comp Cams Xtreme Energy hydraulic cam we installed.

	Advertised Duration (degrees at 0.006-inch)	Duration (degrees at 0.050-inch)	Lift (inches w/1.5 ratio)	Lobe Separation Angle (degrees)
Stock Cam				
Intake	260	194	0.383	112
Exhaust	268	202	0.401	—
Comp Cams				
Intake	268	224	0.477	110
Exhaust	280	230	0.480	—

PORT TESTING

While all the theory supports these efforts, the only way to know for sure is to test these stock heads before and after and report on the results. Typically, pocket porting is worth around 20 hp, with most of the power gains seen above 4,000 rpm. While we did improve the torque, as usual the bulk of the improvements occurred above the 4,000 rpm mark. Keep in mind that all we did for this first test was yank the heads off, do the pocket porting, and stuff them back on the Goodwrench grist mill. From last chapter's best numbers of 265 hp at 4,400 and 350 lb-ft of torque at 3,600, the combination of the pocket-porting work and the compression ratio increase pushed our Good-wrench Quest 350 to a solid increase of 26 hp at 5,300 rpm. The cork was now the miniscule camshaft. It was time for a hotter cam.

The stock cam was downright wimpy, (see the Cam Specs Chart) and now that we had a set of cylinder heads that were up to the task, it was clearly time to take advantage of the heads with more duration and lift. There's always a tradeoff when swapping camshafts. The stock cam produced a smooth idle but didn't make great power. The Comp Cams Xtreme Energy 268H cam we chose bumped the duration at 0.050 inch by a massive 30 degrees on the intake side and 28 degrees on the exhaust. Even more amazing is the lift comparison, where the intake valves now jump 0.094 inch higher and the exhaust gains 0.079 inch!

After we installed the cam, we turned Duttweiler loose on the dyno and pushed the Goodwrench 350 to a thumpin' 336 hp at 5,300 rpm, with peak torque jumping up to a killer 377 lb-ft at 3,800 rpm. These are excellent numbers for a basic 350 with a few simple modifications.

CONCLUSION

Are we satisfied with the power? Hardly. We've pumped it up from a measly 236 hp to a righteous 336—a massive 100 hp over the bone-stock baseline. Next, we're all set to try a set of Corvette L98 aluminum heads from GM Performance Parts, and then we're going to pocket port them to see what they can do. After that, we've got a set of those trick iron Vortec heads to try. And then maybe a bigger camshaft and then…well, even we haven't gotten that far yet.

Small-Block Chevy Engine Buildups

Dyno Test Results

Test 1: This is the final test from the last chapter, with the Goodwrench 350 fitted with a set of 1 5/8-inch Hooker headers and an Edelbrock Performer dual-plane intake manifold.

Test 2: Same as Test 1 but with pocket-ported cylinder heads and a thinner head gasket for more compression.

Test 3: Same as Test 2 with the addition of the Comp Cams 268 Xtreme Energy cam.

Test 4: Not all dyno tests run the way you want them to. After completing the tests of the heads and the camshaft at a 300-rpm-per-second acceleration rate on the dyno, we noticed the engine lost serious torque below 3,500 rpm. We ran another test at a much slower acceleration rate and the torque losses at this point became major increases over Test 1. In the car, this would have been obvious as a bog. We traced the torque loss to an air valve door adjustment. The slower-rate dyno test revealed massive torque improvements that are more indicative of what this engine is capable of producing with the better cam. This indicates just how critical fuel curve tuning is to make good power.

RPM	TEST 1 TQ	TEST 1 HP	TEST 2 TQ	TEST 2 HP	DIFF TQ	DIFF HP	TEST 3 TQ	TEST 3 HP	DIFF TQ	DIFF HP	TEST 4 TQ	TEST 4 HP
2,500	255	121	261	124	6	3	261	124	6	3	—	—
2,600	267	132	268	132	1	—	262	129	-5	-3	—	—
2,700	265	136	271	139	6	3	264	136	-1	—	—	—
2,800	274	146	276	147	2	1	264	141	-10	-5	—	—
2,900	274	151	279	154	5	3	264	146	-10	-5	—	—
3,000	274	157	285	163	11	6	257	147	-17	-10	93	52
3,100	277	163	292	172	15	9	255	151	-22	-12	91	54
3,200	277	168	302	184	25	16	272	166	-5	-2	98	61
3,300	291	183	308	193	17	10	281	177	-10	-6	88	55
3,400	330	214	337	218	7	4	293	190	-37	-24	50	32
3,500	347	231	358	239	11	8	358	239	11	8	34	23
3,600	350*	240	362	248	12	8	373	256	23	16	—	—
3,700	350	246	361*	255	11	9	376	265	26	19		
3,800	347	251	360	260	13	9	377*	272	30	21		
3,900	344	256	358	266	14	10	375	279	31	23		
4,000	338	257	354	270	16	13	373	284	35	27		
4,100	333	260	350	273	17	13	370	289	37	29		
4,200	329	263	346	277	17	14	370	296	41	36		
4,300	323	265*	340	279	17	14	370	303	47	38		
4,400	315	264	334	280*	19	16	370	310	55	46		
4,500	306	262	326	279	20	17	368	315	62	53		
4,600	296	259	318	278	22	19	365	319	69	60		
4,700	288	258	309	276	21	18	361	323	73	65		
4,800	277	253	299	273	22	20	356	325	79	72		
4,900	267	250	290	271	23	21	352	329	85	79		
5,000	259	247	282	269	23	22	350	333	91	86		
5,100	250	242	274	266	24	24	345	335	95	93		
5,200	240	237	266	263	26	26	338	334	98	97		
5,300	231	233	257	259	26	26	333	336*	102	103		

The Goodwrench Quest, Part 3

By Jeff Smith
Photography by Ed Taylor

4

In the continuing saga of the Goodwrench Quest, we now embark on a journey into the exotic world of aluminum heads. In this chapter, we'll start with a set of affordable GM Corvette L-98 aluminum cylinder heads from GM Performance Parts to see what power these heads can deliver. Last month, we pocket-ported the iron heads while retaining the stock 1.94-/1.50-inch valve sizes. We also bolted in a much larger Comp Cams Xtreme Energy dual-pattern, flat-tappet hydraulic cam that, together with the Performer intake and Hooker headers, was worth a total of 336 hp at 5,300 rpm and torque of 377 lb-ft at 3,800. The original baseline with headers and the Performer intake was a meager 265 hp and 350 lb-ft of torque. We've already made great gains, but we're not through yet.

The Corvette aluminum heads are a great bolt-on for mild street engines for a number of reasons. They are some of the least expensive aluminum heads on the market, with a new pair selling for roughly $800, complete. While the small 58cc combustion chamber size can be a compression problem for 350s, in this case the small chamber puts the squeeze right where you want it. With the 73cc iron heads, compression on the Goodwrench 350 was barely 8.4:1. Just bolting on the Corvette heads increased compression to 10.1:1, which should be worth some power

yet it will still tolerate 92-octane pump gas.

THE TEST

For the sake of continuity, Test 1 in our dyno charts is the best power combination from last month, with the pocket-ported stock iron heads combined with the Edelbrock Performer intake and the Q-jet. This became the new baseline. Test 2 added only the stock aluminum Corvette heads with everything else remaining the same. Given that the airflow figures between the stock aluminum heads and the pocket-ported iron heads are so similar, this gives us a chance to see what gains can be had from the increased compression. With this done, we decided that Test 3 would evaluate the induction system change to an Edelbrock Performer RPM intake and a Holley 750-cfm carburetor. This gave us a chance to see if the manifold would add a little more power. Finally, Test 4 looks at whether any advantage would be gained by pocket-porting the Vette heads.

With the Goodwrench 350 back on Ken Duttweiler's dyno, Test 2 started with the stock Corvette aluminum heads. As soon as the engine fired up and settled into an idle, it was obvious

Small-Block Chevy Engine Buildups

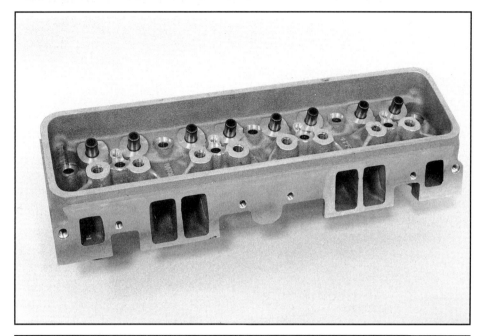

In this chapter we wanted to evaluate the performance advantage of a set of Corvette aluminum heads. These heads were originally used on L98 Corvette Tuned Port Injected (TPI) engines that preceded the LT1 engine. The heads are extremely lightweight with small ports and require a centerbolt valve cover configuration. If you choose to use roller rockers with these heads, there may be some interference with the centerbolt fixtures and some machining of the covers (especially the Corvette magnesium style) may be required.

bump in compression.

Test 3 left everything the same except the induction system. Ed added the Performer RPM intake and Holley 750-cfm carburetor, with the intention that we would mainly be evaluating the effect of the intake manifold and carb on the engine. Since this Goodwrench engine was nowhere near the maximum airflow limit of the Q-jet, the reason for the change was mainly because most enthusiasts would have converted over to a Holley carburetor. The Holley was used with out-of-the-box jetting, and timing remained at 36 degrees total lead. From the power chart you can see that the intake produced a significant gain in torque virtually through the entire powerband, with a torque gain of an amazing 25 lb-ft at 3,500 rpm where peak toque occurred. This improvement is probably evenly split between the better manifold and the Holley's more stable fuel curve. Peak horsepower increased slightly from 348 to 355 at almost the same rpm.

Now it was time to see if some pocket-porting work on the aluminum heads would pay off with a power dividend. The airflow numbers hinted that the power should improve, especially on the higher rpm side since the Vette heads gained much more exhaust flow than intake. The dyno flogging in Test 4 pointed in that direction, but at a cost. Note that between 3,100 and 3,900 the torque dropped off compared to Test 3. This is probably due to the over-efficient exhaust port combined with the cam's longer duration exhaust lobe. For this particular application, we would have been better served to install a single pattern cam with a wider lobe separation angle to maximize the flow

that we had increased compression because the engine now had a decided lope to the idle that had not been noticeable with the iron heads. Once warmed up, CHP's intrepid dyno man Ed Taylor yanked the handle and made a series of pulls to determine timing, uncovering best power numbers of 348 hp at 5,500 rpm with max torque coming in at 380 lb-ft at 3,600 rpm. This included the use of the Performer intake and Q-jet along with the 1 5/8-inch Hooker headers and a pair of 2 1/4-inch mufflers. Power didn't really increase over the ported iron heads, even with the 1.7:1

The Goodwrench Quest, Part 3

The 58cc chamber on the L98 heads is substantially smaller than the typical 76cc iron cylinder heads, which drastically increases compression. Since our Goodwrench engine was seriously down on compression to begin with, the squeeze factor went from a measly 8.4:1 (using a 0.015-inch Fel-Pro head gasket) to a much more agreeable 10.1:1 with the smaller chamber.

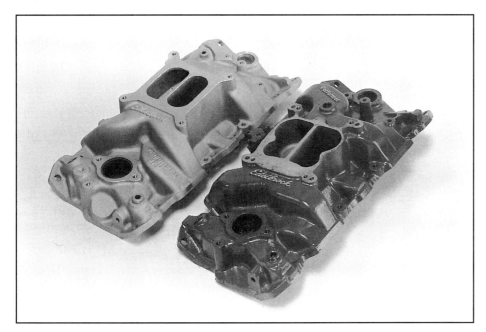

The Edelbrock Performer RPM (left) is the big brother to the more stock-appearing Performer. Both are dual-plane designs, but the RPM uses a taller carburetor mounting flange with larger runners to flow more air while still creating excellent torque for street use. As you can see from our tests, even with stock Corvette heads, the intake and carb combination is worth as much as 25 lb-ft of torque at 3,500.

numbers. This is why we always stress looking at the entire power curve rather than just the peak numbers. While Test 4's peak torque was almost identical at 401 lb-ft and horsepower improved slightly with 361 hp at 5,500 over the stock heads at 355 hp, the loss of torque in the midrange would result in minimal (if any) improvement on the dragstrip.

If we go back to the Test 1 numbers of 366 hp and 377 lb-ft of torque, it's clear that the best overall power curve in this series of tests lies with Test 3 using the stock Vette heads, a Holley 750 carburetor, an Edelbrock Performer RPM, Hooker 1 5/8-inch headers, and a pair of 2 1/4-inch turbo-style mufflers. The torque curve is amazingly strong, with over 390 lb-ft of torque from 3,100 to 3,900, and is never less than 350 lb-ft from 2,500 to 5,200 rpm. This combination also produced over 100 hp and 50 lb-ft of torque more than the original baseline using a stock aluminum intake, a Q-jet, and cast-iron exhaust manifolds. But we have barely scratched the surface of what can be done to this otherwise stock short-block 350 Chevy. In fact, before you rush right out to buy a set of GM Performance Parts aluminum Vette heads, you might want to wait until the next chapter to see what kind of power we can pull out of a set of those excellent GMPP Vortec iron heads.

We won't spoil the surprise by spilling the numbers here, but if ultimate power is what you're seeking from a mild-mannered 350 Chevy, then you'll certainly want to read on into the next few chapters. There's power to be had with these heads. The best news is that these Vortec heads are extremely affordable at less than $400 for a complete pair from Scoggin-Dickey and other GMPP dealers! So if a powerful small-block for less bucks makes your mouth water, move on to the next few chapters.

Small-Block Chevy Engine Buildups

We chose Holley's 4779 750-cfm double-pumper not only because it is an excellent street carburetor but also because it is one of the most popular carburetors on the market. Besides, we wanted to test the Goodwrench small-block with off-the-shelf carburetors that are less expensive than highly modified units.

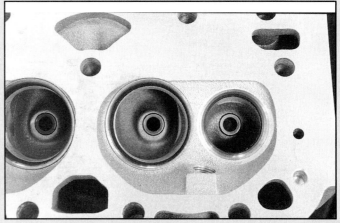

This is the stock Vette chamber. McKenzie again performed a minimal amount of porting within an inch of the valve seat on both ports to duplicate what could be done both easily and inexpensively. As you can see from the numbers in the Flow Testing chart, the exhaust improved dramatically while the intake port did not respond as well to the pocket work.

All late-model engines now use self-aligning rocker arms, which are identified by the "rails" or ribs on the stock stamped rockers that are used to keep them centered over the valve tip since no guideplates are used. Comp Cams offers rail-style Magnum roller-tipped rocker arms that are a direct replacement. We tried increasing the rocker ratio to 1.6:1 in this configuration, with no improvement.

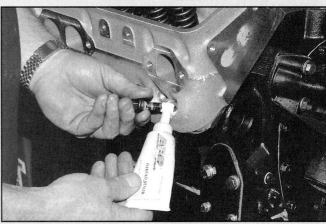

We used ARP head bolts with thread sealant to ensure a positive seal to the block as well as to prevent coolant leakage past the threads.

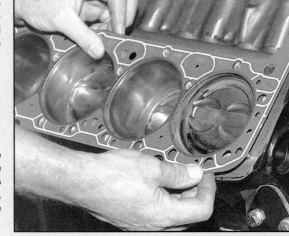

Because of the aluminum heads' small 58cc chamber size, we used Fel-Pro 0.041-inch-thick composition gaskets to keep the compression within the 92-octane pump gas limits. Test 1 versus Test 2 is basically a comparison of the effect of compression, since the pocket-ported iron heads flow very similarly to the stock Corvette heads.

The Goodwrench Quest, Part 3

FLOW TESTING

The following chart compares the pocket-ported iron heads to both the stock and pocket-ported aluminum Corvette heads. As you can see, the ported iron heads flow a little better than the stock aluminum Vette heads. Contrasting the ported versions of the two heads, you can see that the massaged aluminum Vette heads reveal a much stronger exhaust port. The average exhaust-to-intake flow relationship for the L98 Vette heads is a very high 88 percent, with low-lift figures peaking at 94 percent at 0.200-inch valve lift. This means that the engine would have preferred a single-pattern camshaft with a wider lobe separation angle to take maximum advantage of this configuration. All values are given in cfm with the test performed by McKenzie's Cylinder Heads at 28 inches of water on a Superflow 600 flow bench. Note that the stock L98 intake port at 0.500-inch valve lift suffered from turbulence and no flow numbers could be obtained.

Valve Lift	Ported Iron Int.	Stock L98 Int.	Ported L98 Int.	Ported & Back-cut Iron Exh.	Stock L98 Exh.	Ported L98 Exh.
0.100	59	62	59	47	53	53
0.200	119	115	119	103	101	112
0.300	175	159	172	136	134	153
0.400	196	189	212	150	157	175
0.500	197	—	220	160	171	191

POWER CHARTS

The baseline Test 1 in this series uses the best overall power numbers from the last chapter's exercise. This configuration employed pocket-ported stock iron heads, a Comp Cams 268 Xtreme Energy camshaft (224/230 degrees at 0.050 with 0.477/0.480-inch lift and a 110-degree lobe separation angle) along with an Edelbrock Performer intake, a Q-jet, stock HEI, and a set of Hooker 1 5/8-inch headers plumbed into a pair of 2 1/4-inch Hooker turbo mufflers. This established the baseline for Test 2, which tested a set of stock Corvette L98 aluminum heads from GM Performance Parts along with a set of Comp Cams 1.5:1 Magnum roller rockers. Test 3 added a taller Edelbrock Performer RPM intake along with a Holley 750-cfm carburetor. Finally, Test 4 used the same components except for pocket-porting work performed by Todd McKenzie at McKenzie's Cylinder Heads.

The "Increase" column to the far right is the difference in power between Test 1 and Test 3. This is the power gained from the addition of the stock aluminum heads, the Edelbrock Performer RPM intake, and the Holley 750 carburetor. The dashes in Test 1 are data that wasn't gathered.

RPM	TEST 1 TQ	TEST 1 HP	TEST 2 TQ	TEST 2 HP	TEST 3 TQ	TEST 3 HP	TEST 4 TQ	TEST 4 HP	INCREASE (1 vs. 3) TQ	INCREASE (1 vs. 3) HP
2,500	339	161	351	168	350	166	-	-	-	-
2,600	340	168	354	176	344	171	-	-	-	-
2,700	341	175	356	183	362	186	-	-	-	-

continued on next page

Small-Block Chevy Engine Buildups

RPM	TEST 1 TQ	TEST 1 HP	TEST 2 TQ	TEST 2 HP	TEST 3 TQ	TEST 3 HP	TEST 4 TQ	TEST 4 HP	INCREASE (1 vs. 3) TQ	INCREASE (1 vs. 3) HP
2,800	-	-	345	184	360	192	372	198	-	-
2,900	-	-	351	194	368	203	376	208	-	-
3,000	367	209	358	204	381	217	382	218	15	9
3,100	367	217	369	218	393	231	390	230	23	13
3,200	375	228	372	227	396	241	392	239	17	11
3,300	379	238	373	234	398	250	394	247	15	9
3,400	380	246	376	243	400	259	395	256	15	10
3,500	381	254	377	251	402*	268	399	265	18	11
3,600	378	259	380*	260	402	275	401*	274	23	15
3,700	376	265	380	268	400	282	398	281	22	16
3,800	377	272	379	274	399	289	395	286	18	14
3,900	375	279	375	279	394	293	392	291	17	12
4,000	373	284	373	284	389	296	390	297	17	13
4,100	370	289	368	287	387	302	390	305	20	16
4,200	370	296	365	292	385	308	386	309	16	13
4,300	370	303	365	299	384	314	387	317	17	14
4,400	370	310	364	305	383	321	384	321	14	11
4,500	368	315	361	309	383	328	382	328	14	13
4,600	365	319	360	315	378	331	383	335	18	16
4,700	361	323	357	320	374	334	382	341	21	18
4,800	356	325	354	323	369	337	382	349	26*	24*
4,900	352	329	347	324	365	341	378	353	26	24
5,000	350	333	347	331	362	345	371	353	21	20
5,100	345	335	345	336	357	346	361	351	16	16
5,200	338	334	342	339	351	348	354	351	16	17
5,300	333	336	340	343	349	352	350	354	17	18
5,400	-	-	337	347	345	355*	346	355	-	-
5,500	-	-	333	348*	337	353	344	361*	-	-
5,600	-	-	325	347	325	347	336	359	-	-
5,700	-	-	316	343	313	340	324	352	-	-
5,800	-	-	306	338	298	328	311	343	-	-

The Goodwrench Quest, Part 4
By Jeff Smith
Photography by Ed Taylor

In the continuing saga of our Goodwrench budget 350, we're now approaching serious horsepower. Because we know how you think, you've already skipped to the dyno charts, so it's no surprise to tell you we've made 384 hp and an impressive 401 lb-ft of torque all below 6,000 rpm. But this information doesn't do much good unless you know how we did it. Here's the skinny.

As you may remember from the last chapter, we had rigged our Scoggin-Dickey Goodwrench 350 with a set of aluminum L-98 Corvette heads first used on TPI Corvettes. As it turned out, the stock, unported heads made the best overall power, while the ported versions made more peak horsepower but cost torque. Ultimately, a single-pattern camshaft with the same exhaust specs as intake would have worked best in that specific application. That would have improved torque and perhaps even pumped the horsepower.

But now we were ready to test a set of heads that not only have excellent flow numbers but are extremely affordable as well. Generally, rebuilding a set of stock iron heads will cost in the neighborhood of $400 to $500. The problem with this is that all you have to show for five big ones is a marginal set of stock heads. A better solution is the new Vortec iron factory cylinder head first used on the '96 Vortec L31 Chevrolet pickup engines. This is an amazing 1.94/1.50-inch

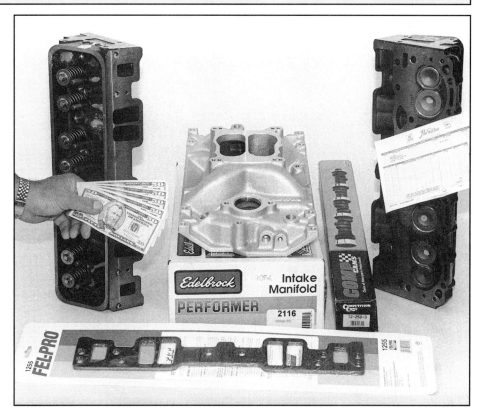

valve head that not only outflows a stock cast-iron Bow Tie head but also is hugely affordable. Scoggin-Dickey will sell you a pair of these heads for $399 per set, complete with valves, springs, retainers, keepers, and rocker studs. That's an awesome deal. But there's more to the story.

THE PLAN

The Vortec heads use a different intake manifold pattern than a stock small-block Chevy cylinder head. While it is possible to drill the heads for a standard small-block six-bolt intake pattern, it's also possible to ruin a head if drilled incorrectly. A better approach is to use one of the new intake manifolds from Edelbrock that are designed to fit the new Vortec heads. To this end, we used an Edelbrock Performer RPM intake and an out-of-the-box Holley 0-4779 750-cfm carburetor to complete the induction package. We also retained the Hooker 1 5/8-inch street headers and the 2 1/4-inch Hooker-muffled exhaust system.

THE HEADS

So why are these production heads so good? Chevy borrowed the excellent intake and exhaust port design from the LT1 aluminum head and

Small-Block Chevy Engine Buildups

We stuck the Goodwrench 350 back on the dyno with the Corvette-style valve covers on it (the Vortec heads require centerbolt valve covers) and twisted it up until it cranked out an amazing 384 hp at 5,700 rpm. With its great torque curve, this is an excellent street combination that's just a whisker away from 400 hp.

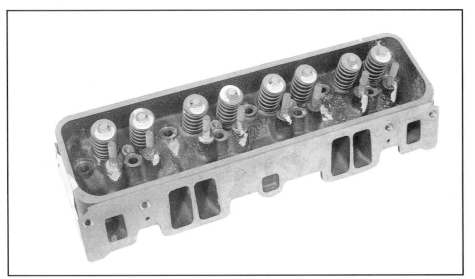

This is the cylinder head that's causing all the excitement. The Vortec iron small-block head is nothing more than a stock L-31 iron truck head. But Chevy learned from the success of its aluminum Corvette LT1 and duplicated the LT1 ports in the Vortec head. The Vortec heads come complete with valves, springs, and press-in studs.

McKenzie of McKenzie's Cylinder Heads installed the appropriate Comp Cams valvesprings and new seals on the heads to allow the cam to do its stuff without suffering from valve float. While pushrod guideplates could have been installed, we already had a set of rail-style rocker arms from Comp, so we elected to stick with these and avoid the added cost of machining the heads for screw-in studs and guideplates.

While these heads offer excellent flow potential (see "Flow Chart" sidebar), power-crazed hot rodders always want to make them better. McKenzie's previous experience with these heads revealed that overzealous pocket-porting can hurt flow rather than improve it. So, to duplicate what a typical hot rodder might do, the heads were first tested stone stock on the dyno, then removed and just lightly dusted with the grinder to remove the lip just below the factory three-angle valve job to improve low-lift flow. Then Todd added a 30-degree back-cut to the exhaust valves that bumped up the flow below 0.250 inch of lift.

All this is conjecture, however. The real test is bolting the parts on and twisting the motor on Ken Duttweiler's dyno to see what happens. The numbers don't lie. These budget iron heads are probably the best per-dollar deal on the planet.

THE TEST

If you refer to the Power Chart that lists the dyno test results, we're using Test 1 as the baseline, which was the best overall power curve from the test of the Corvette alu-minum heads. At 355 hp at 5,400 and 402 lb-ft of torque at 3,500, this is an excellent street package. One point worth

dropped it into the iron Vortec head, which includes a small but efficient 64cc combustion chamber and pressed-in 3/8-inch studs. The stock springs are a little weak for the kind of abuse we intended, so Todd

The Goodwrench Quest, Part 4

While the Vortec head does bolt on to any small-block Chevy (except the LT1, LT4 or LS1), it has a different intake manifold bolt pattern that uses only four intake bolts per side, for a total of eight. Also, the bolt holes are drilled at a steeper angle. While it is possible to either modify an existing intake manifold or drill and tap the head to accept a standard small-block intake, the easiest route is to go with a dedicated Vortec intake from Edelbrock.

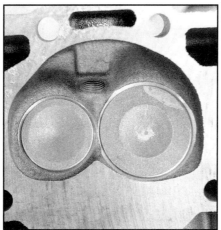 

These photos compare the Vortec combustion chamber (left) with a typical '70s chamber from a production 441 head (right). Note how the Vortec head is more kidney-shaped, which is an attempt to improve combustion efficiency by inducing swirl in the chamber. The other major difference in the Vortec head is a 64cc chamber while the older 441 head has a 76cc chamber.

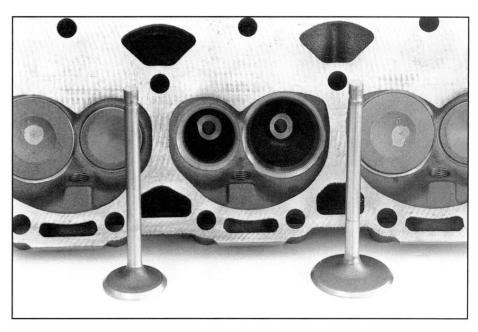

Valve sizes for the Vortec head are the standard 1.94/1.50-inch variety. McKenzie performed a very minor amount of port work below the valve seats of both the intake and exhausts and also put a 30-degree back-cut on the exhaust valve. Even this minimal amount of work was worth 13 hp.

noting is that adding the stock iron Vortec heads on the Goodwrench 350 also increased the compression ratio. The Vortec heads come with a 64cc combustion chamber, while the Corvette heads use a smaller 58cc chamber. However, this 64cc chamber is still smaller than the stock Goodwrench 76cc iron castings. This means that the compression is down from the aluminum heads but is still higher than the stock iron heads.

Looking at the results from Test 2, you can see that the Vortec suffers slightly compared to the Corvette heads below 4,000 rpm, but this is an average power loss of barely 4 lb-ft of torque, which would be tough to detect in the car. The good news is that above 4,000 rpm, the heads cranked an average increase of 13 hp from 4,000 to 5,800 rpm with a maximum increase of 38 hp at 5,800. That's power you can feel!

For Test 3, McKenzie put about a half-hour into a minor tune-up of the area directly below the valve seat in the intake and exhaust ports and machined a 30-degree back-cut on the stock exhaust valves to improve low-lift flow. Ed Taylor then bolted the heads back on the engine and tested it again, with no other changes. Again, the engine suffered in the torque comparison versus the Corvette heads below 4,000 rpm, but the total average loss was less than 2 lb-ft of torque. On the plus side, above 4,000 rpm the porting had a chance to strut its stuff. While peak horsepower jumped slightly to a best of 384 hp at 5,700, the interesting info was that the entire torque curve above 4,000 averaged an increase of over 20 lb-ft at each rpm point. Peak torque didn't change significantly at 401 lb-ft at

3,600. From this test, you can imagine this motor would really pull between 4,000 and 6,000 rpm.

CONCLUSION

This put our under-$2,500 pedestrian cast piston Goodwrench 350 at a stout 384 hp at 5,700 rpm and only an agonizing 16 hp away from 400. This is so tantalizingly close that we're not quitting yet. We think we can break the 400 hp barrier with this engine without resorting to exotic aftermarket parts. The plan involves some minor tuning, perhaps a bigger set of headers, or different mufflers, but you'll have to move on to the next chapter to find out if we pulled it off.

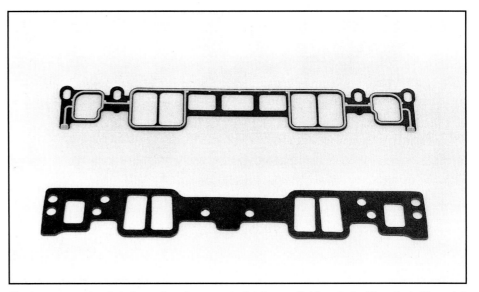

You have a couple of choices when it comes to intake manifold gaskets. The stock Chevrolet gasket (top) offers snap-in-place convenience with a silicone outline around the ports and the water jackets. Fel-Pro also offers a replacement gasket (bottom) that will work with either the 8-bolt Vortec pattern or the standard small-block Chevy 12-bolt intake manifold pattern. This is a Vortec-specific gasket because the port layout on the Vortec head is wider at the bottom than it is at the top, which would require trimming on a standard small-block gasket.

Edelbrock currently has three different intake manifolds designed to bolt directly to the Vortec cylinder head. We chose the Edelbrock Performer RPM Vortec that accepts the square-pattern Holley carburetor flange. The intake in this photo is actually the standard Performer that will accept either a Q-jet or a Holley. The carburetor we chose is a Holley 0-4779 750-cfm double-pumper

FLOW CHART

The following chart compares the ported L98 Corvette aluminum head to the iron Vortec head stock, pocket-ported, and ported with a 30-degree back-cut on the exhaust side only. As you can see, the combination of all three creates a very efficient exhaust port. Like the Corvette heads, this may have contributed to the minor torque loss observed on Tests 2 and 3. One way to overcome this power loss would be to choose a single-pattern camshaft. All figures are given in cfm, with all tests performed on McKenzie's SuperFlow flow bench at 28 inches of water test depression.

Valve Lift (inches)	Ported Aluminum Vette		Stock Vortec		Ported Vortec		Ported & Back-cut Vortec
	I	E	I	E	I	E	E
0.100	059	053	059	050	060	050	062
0.200	119	112	118	111	128	101	119
0.300	172	153	177	148	193	161	162
0.400	212	175	217	160	224	181	181
0.500	220	191	226	164	239	190	190

PART NUMBERS

The following are part numbers for both GM and aftermarket parts used on the Goodwrench 350.

Component	Source	Part Number
Goodwrench 350	Scoggin-Dickey	10067353
HEI distributor	GMPP	1104067
Vortec heads, complete	GMPP	12558060
Intake gasket, Vortec	GMPP	12529094
Corvette heads, L98	GMPP	12556463
Intake, Performer	Edelbrock	2116
Intake, Performer RPM	Edelbrock	7116
Intake, Super Victor	Edelbrock	2913
Intake gasket, Vortec	Fel-Pro	1255
Head gasket, 0.041-inch	Fel-Pro	1003
Camshaft	Comp Cams	XE-268H-10
Head bolts	ARP	134-3601
Perma-Loc adjusters	ARP	300-8241
Headers, 1 5/8-inch	Hooker	Application specific
Mufflers, 2 1/4-inch	Hooker	21005
Carburetor, 750-cfm	Holley	0-4779

POWER CHART

The following chart shows the power difference between the stock Corvette aluminum heads in Test 1 versus the stock Vortec iron heads in Test 2. Test 3 reveals the power difference between the pocket-ported Vortec heads compared to Test 1. Notice how the Vortec heads lose a small amount of torque up to 3,800 compared to the aluminum Vette heads. But from 4,000 rpm up, the power advantage swings mightily over to the Vortec heads. While peak horsepower can be misleading, it's hard to ignore a 51hp gain at 5,800 rpm.

RPM	TEST 1 TQ	TEST 1 HP	TEST 2 TQ	TEST 2 HP	1 v. 2 TQ	1 v. 2 HP	TEST 3 TQ	TEST 3 HP	1 vs. 3 TQ	1 vs. 3 HP
2,500	351	168	345	165	-6	-3	345	165	-6	-3
2,600	354	176	347	172	-7	-4	345	171	-9	-5
2,700	356	183	350	180	-6	-3	351	180	-5	-3
2,800	360	192	357	190	-3	-2	358	191	-2	-2
2,900	368	203	375	207	-7	-4	379	209	+11	+6
3,000	381	217	381	217	–	–	385	220	+4	+3
3,100	393	231	384	227	-7	-4	387	228	-6	-3
3,200	396	241	391	238	-5	-3	392	239	-4	-2
3,300	398	250	394	247	-4	-3	398	250	–	–
3,400	400	259	395	256	-5	-3	398	258	-2	-2
3,500	402*	268	400*	266	-2	-2	400	267	-2	-2
3,600	402	275	398	273	-4	-2	401*	275	-1	
3,700	400	282	397	280	-3	-2	397	280	-3	-2
3,800	399	289	396	287	-3	-2	398	288	-1	-1
3,900	394	293	394	293	–	–	397	295	+3	+2
4,000	389	296	395	301	+6	+5	395	300	+6	+4
4,100	387	302	397	310	+10	+8	397	310	+10	+8
4,200	385	308	397	317	+8	+9	396	317	+11	+9
4,300	384	314	394	322	+10	+8	397	325	+13	+11
4,400	383	321	390	327	+7	+6	399	334	+16	+13
4,500	383	328	389	333	+6	+5	395	339	+12	+11
4,600	378	331	388	340	+10	+9	392	343	+14	+12
4,700	374	334	387	346	+13	+12	390	349	+16	+15
4,800	369	337	382	349	+13	+12	388	355	+19	+18
4,900	365	341	380	354	+15	+14	386	360	+21	+19
5,000	362	345	374	356	+12	+11	384	366	+22	+21
5,100	357	346	367	357	+10	+11	381	370	+24	+24
5,200	351	348	363	359	+12	+11	374	370	+23	+22
5,300	349	352	361	364	+12	+12	368	371	+19	+19
5,400	345	355*	358	368	+13	+13	366	377	+21	+22
5,500	337	353	352	368	+15	+15	363	380	+26	+27
5,600	325	347	346	369	+21	+22	359	383	+34	+36
5,700	313	340	342	371*	+29	+31	354	384*	+31	+44
5,800	298	328	331	366	+33	+38	343	379	+45	+51

The Goodwrench Quest, Part 5

By Jeff Smith
Photography by Ed Taylor

In the continuing saga of the Scoggin-Dickey–supplied Goodwrench 350 that we've been beating on for four chapters, the story is really starting to get interesting. Here are the details: 408 hp at 5,800 and a stunning 430 lb-ft of torque at 3,700 rpm with only minor changes to the engine! How did we do it? Just follow the bouncing ball.

In the last chapter, we bolted on a set of GM Performance Parts' iron Vortec heads that flow air like a wind tunnel. We achieved 384 hp by simply pocket-porting the heads and bolting them to our existing combination. As a quick overview, the engine is a stock bottom-end GM Goodwrench 350 supplied by Scoggin-Dickey with a Comp Cams 268 Xtreme Energy camshaft (224/230 degrees at 0.050-inch tappet lift and 0.477/0.480-inch valve lift with 110-degree lobe separation angle), roller-tipped rockers, and the Vortec heads. The rest of the engine is outfitted with an Edelbrock Performer RPM intake, a Holley 750 double-pumper carburetor, Hooker 1 5/8-inch headers, and a pair of Hooker 2 1/4-inch turbo mufflers. That's it, except for the stock HEI distributor, MSD wires, and 92-octane pump gas.

TESTING, TESTING

After the last experiment with the iron heads, we really wanted to make 400 hp. While we could have added a bigger cam, a single-plane intake, or

With the Goodwrench 350 bolted back on Ken Duttweiler's dyno, we wanted to see 400 hp. What surprised us wasn't that we made 408 hp but that this mild-mannered 350 could smoke to the tune of 430 lb-ft of torque.

even nitrous, we wanted to see if we could hit four big ones without major changes. In all previous stories, we merely bolted parts on the engine and ran the tests. In virtually every case, we could have taken the time to fine-tune each change by running through the ignition timing and jetting exercise. We didn't do that because we decided to perform these tests by just bolting the parts on. But now that we were so close, we felt that a few tune-ups might push us over the top.

Ed Taylor has ramrodded this buildup from the beginning, including doing the parts-swapping and the dyno-testing. His first tune-up idea was the best. In the previous test, we had used a set of Fel-Pro composition head gaskets that put the engine at 9.1:1 compression (we got that wrong earlier—math was never our strong suit). We knew we could get away with a little more compression, so Ed swapped the composition gaskets for thinner rubber-coated Fel-Pro gaskets. This bumped the com-pression to 9.4:1, which is a safe com-pression ratio for iron heads and 92-octane fuel. A point in our favor to defuse the detonation problem was the engine only wanted 34 degrees of total timing for the best power.

The first pull on the dyno with the added compression was encouraging when torque improved throughout the

Small-Block Chevy Engine Buildups

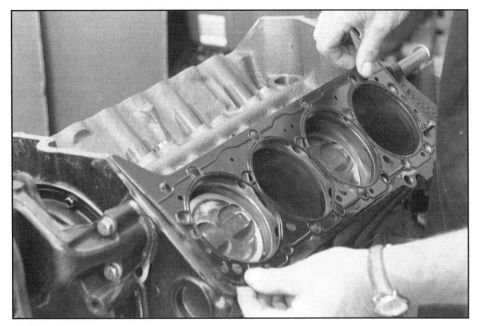

One of the biggest gains was increasing compression by using the Fel-Pro 0.015-inch-thick, rubber-coated shim gaskets. This bumped the compression from 9.1:1 to 9.4:1 and was worth power throughout the entire rpm range.

During an early part of this month's testing, the engine died and wouldn't refire. After some quick diagnostic work, we discovered the stock module in the HEI had failed. A quick replacement from Pep Boys put us back on track.

One swap that did result in more power was a set of Comp Cams roller-tipped Magnum 1.6:1 rockers that replaced the 1.5:1 rockers used earlier. The performance gain was as much as 11 lb-ft of torque and 9 hp. This is attributable to the fact that the heads flow very well at around 0.500- to 0.510-inch lift, which is max lift with the 1.6:1 rockers.

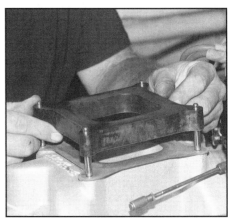

Among the pieces we tried that didn't work was this phenolic 1-inch spacer that should have increased top-end power but instead resulted in a massive power wheeze throughout the entire rpm band. The results were so bad, we didn't even try to tune around it. In our application, it's obvious the Performer RPM intake didn't like the spacer.

entire power curve, generating 6 to 13 lb-ft of additional torque between 3,500 and 5,100 rpm. Everything else remained the same from the previous test—timing and jetting were also the same with 34 degrees of total timing and stock jetting in the 750 Holley.

The next couple of tests weren't as successful, but they're worth reporting if for no other reason than to save you the effort. We thought a 1-inch open spacer under the carburetor would help top-end power—it didn't. We expected to lose a little torque below the 3,800-rpm peak but nothing like what happened. The motor fell flat on its face, losing not only a massive 57 lb-ft of torque at 4,100 but also an average of 28 lb-ft of torque across the entire powerband! We didn't try adding fuel to improve the situation, even though fuel flow dropped. Basically, the response was so bad we felt that we were flogging a dead horse.

The Goodwrench Quest, Part 5

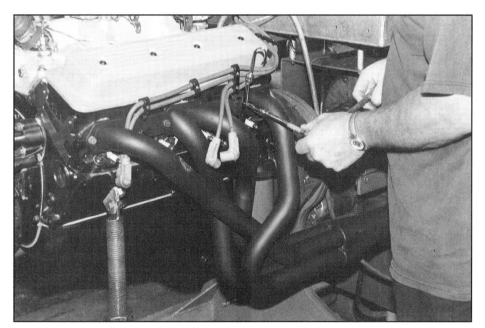

Dyno dueler Ed Taylor also tried a set of 1 3/4-inch Hooker chassis headers along with a pair of Borla mufflers. These headers allow the engine to make more power at high rpm, but we were surprised to see the engine respond with more torque even down at the very bottom rpm levels.

The Borla XR-1 stainless steel mufflers also used 2 1/4-inch inlet and outlet sizes compared to the Hooker turbo-style mufflers tested previously. The Borlas definitely helped power but are also a bit more expensive than the Hookers. It's hard to find fault with 430 lb-ft of torque and 408 hp!

After retesting without the spacer to ensure nothing had broken on the engine, we next tried adding timing to the engine. This also resulted in a loss of power with no substantial gains even at the lower engine speed, where we thought this might help. This leads us to believe that the excellent Vortec combustion chamber design is responsible for the reduced timing.

After only limited success with compression and still 6 hp shy of our 400hp goal, it was time to try something else. Since the Vortec heads flow well in excess of 0.500-inch lift, we thought that more valve lift might help power. We bolted on a set of Comp Cams' Magnum rail 1.6:1 roller-tipped rockers in search of more lift and were rewarded with more torque and horsepower. In fact, for the first time, the Goodwrench 350 exceeded the 400hp watermark with 402 hp at 5,800. More surprising was the torque increase, with a stunning peak torque of 416 lb-ft at 3,600 rpm and increases of well over 10 lb-ft at various points. Now we were really cookin'.

We could have stopped here, but we wanted more. Ed next wanted to try a set of 1 3/4-inch headers to pump up the top-end power, along with a pair of Borla XR-1 stainless steel straight-through-style mufflers. The exhaust combo was worth 6 hp and as much as 13 lb-ft of torque, creating a peak horsepower of 408 at 5,800 and an amazing 430 lb-ft of torque at 3,700. This is outstanding power, especially when you consider that we're working with a basic short-block with probably the worst cast piston design known to small-block Chevys.

When we started this engine test sequence, we harbored aspirations of making 350 to perhaps 375 hp and maybe the same amount of torque. We really didn't think this engine could make over 400 hp and 430 lb-ft of torque. That's truly awesome power from an engine that is docile enough to drive on a daily basis. While we used the Goodwrench engine as the test bed for this experiment, there are several different ways to get to this same place. The easiest step is to buy a set of Vortec heads, a Comp Cams cam, Edelbrock intake, and a set of headers and bolt them on your existing 350 short-block. Depending upon how serious you are about duplicating our combination, it's certainly possible to equal 400 hp.

You could also use a rebuilt short-block with typical four-valve relief pistons (cast or forged) and use the same bolt-on parts we used while investing less time in building the whole engine. We've also bench-raced about substituting the next-step-smaller Comp Cams Xtreme Energy 262 cam to replace the 268. Comp's own testing shows that with stock heads, the 262 cam will make the same power and slightly more torque. This would be a more docile street combination even with the Vortec heads, and it could make even more torque at the cost of perhaps 5 to 10 hp at the peak. This would be the combination we would go for if this was a daily-driven engine.

Small-Block Chevy Engine Buildups

We also included a Summit fuel line with the Holley 750 to make jetting and parts-swapping easier. The line comes with an integrated fuel pressure gauge for quick pressure checks.

CONCLUSION

This Scoggin-Dickey Goodwrench 350 has flat amazed us. With only a few exceptions, virtually everything we've bolted on has improved power, and it's been dead-on reliable. To put this entire project in perspective, we went back to Part 1 to see how far we've come.

Way back then, the out-of-the-box Goodwrench 350 with iron exhaust manifolds, a stock Chevy Q-jet aluminum intake, and a Q-jet carb made 239 hp and a respectable 324 lb-ft of torque at 3,700. Adding the Hooker 1 5/8-inch headers and an Edelbrock Performer but still using the Q-jet, the engine perked up to 265 hp and 350 lb-ft of torque. From the original baseline, we managed to crank this pedestrian Mouse motor with 169 hp while adding 106 lb-ft of torque. That's not bad for a few bolt-on parts. This is especially exciting when you consider that anyone could duplicate our efforts with the Goodwrench 350 for an affordable price.

So have we completed our Goodwrench Quest? Not exactly. This motor has been so successful that now we want to see what it would do with a supercharger bolted on top. Weiand has a very affordable bolt-on supercharger that would look great sitting on top of our Goodwrench motor. This means we'll have to go back to the stock iron heads, since the Weiand blower won't bolt up to the Vortec intake pattern. Stay tuned to see what this rascal does with a blower.

PART NUMBERS

The following are part numbers for both GM and aftermarket parts used on the Goodwrench 350.

COMPONENT	SOURCE	PART NUMBER
Goodwrench 350	Scoggin-Dickey	10067353
HEI distributor	GMPP	1104067
Vortec heads,	GMPP	12558060
Intake gasket, Vortec	GMPP	12529094
Intake, Performer RPM	Edelbrock	7116
Intake gasket, Vortec	Fel-Pro	1255
Head gasket, 0.015"	Fel-Pro	1094
Camshaft	Comp Cams	XE-268H-10
Roller rockers, 1.6:1	Comp Cams	1418-16
Head bolts	ARP	134-3601
Perma-Loc adjusters	ARP	300-8241
Headers, 1 3/4-inch	Hooker	App. specific
Mufflers, XR-1	Borla	40646
Carburetor, 750 cfm	Holley	0-4779

The Goodwrench Quest, Part 5

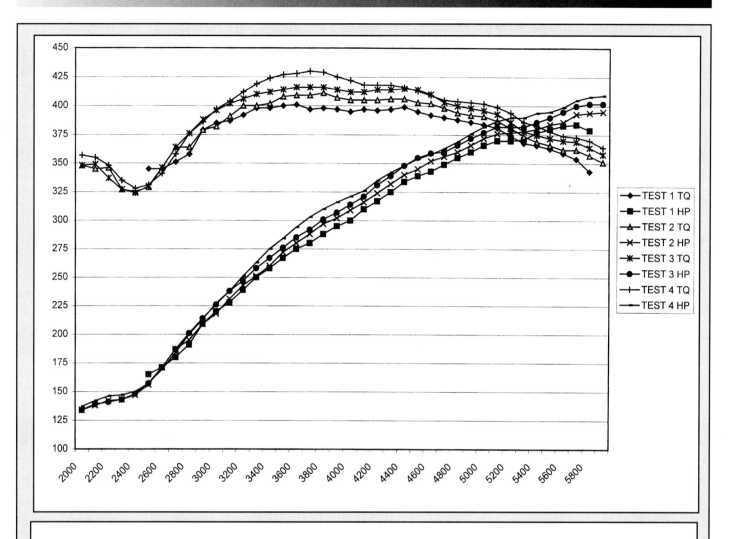

POWER CHART

The following chart uses the best overall power pull as the baseline (Test 1). Test 2 indicates the power increase resulting from the increase in compression. Test 3 added the 1.6:1 roller-tipped rocker arms. As you can see from the numbers and the graph, there is a little power to be gained by opening the valves a little quicker with the increased ratio. Test 4 includes changes from Test 2 and 3 plus a set of 1 3/4-inch Hooker headers and a pair of 2 1/4-inch XR-1 Borla mufflers. As you can see, with just these few changes the power increased by as much as 29 hp while the torque increased by as much as 33 lb-ft. Those are stout gains!

RPM	TEST 1		TEST 2		TEST 3		TEST 4		DIFFERENCE (1 vs. 4)	
	TQ	HP	TQ	HP	TQ	HP	TQ	HP	TQ	HP
2,000	–	–	348	134	348	134	357	137	–	–
2,100	–	–	345	138	349	139	355	142	–	–
2,200	–	–	346	142	337	141	348	146	–	–
2,300	–	–	327	143	327	143	335	147	–	–
2,400	–	–	325	147	324	148	328	150	–	–
2,500	345	165	329	156	330	157	331	157	-14	-8
2,600	345	171	346	171	346	171	341	169	-4	-2
2,700	351	180	364	187	364	187	358	184	+7	+4
2,800	358	191	364	195	376	201	376	200	+18	+9

Small-Block Chevy Engine Buildups

RPM	TEST 1		TEST 2		TEST 3		TEST 4		DIFFERENCE (1 vs. 4)	
	TQ	HP	TQ	HP	TQ	HP	TQ	HP	TQ	HP
2,900	379	209	379	209	388	214	386	213	+7	+4
3,000	385	220	382	218	396	226	397	227	+11	+6
3,100	387	228	391	231	402	238	404	238	+17	+10
3,200	392	239	400	243	406	247	412	251	+20	+12
3,300	398	250	400	251	410	258	419	263	+21	+13
3,400	398	258	402	260	412	267	424	275	+26	+17
3,500	400	267	408	272	414	276	427	284	+27	+17
3,600	401*	275	409	280	416	285	428	294	+27	+19
3,700	397	280	409	288	416	292	430	303	+33	+23
3,800	398	288	411	297	416	301	429	310	+31	+22
3,900	397	295	407	302	414	307	425	316	+28	+21
4,000	395	300	405	309	412	314	422	321	+27	+21
4,100	397	310	405	316	412	321	418	326	+21	+16
4,200	396	317	405	324	414	331	418	335	+22	+18
4,300	397	325	406	332	414	339	418	342	+21	+17
4,400	399	334	406	340	415	348	416	348	+17	+14
4,500	395	339	403	345	414	355	413	354	+18	+15
4,600	392	343	402	352	410	359	409	358	+17	+15
4,700	390	349	398	356	403	360	405	363	+15	+14
4,800	388	355	394	360	400	366	404	369	+16	+14
4,900	386	360	392	366	398	372	403	376	+17	+16
5,000	384	366	391	372	396	377	402	382	+18	+16
5,100	381	370	387	376	393	382	399	387	+18	+17
5,200	374	370	381	377	386	382	394	390	+20	+20
5,300	368	371	374	378	378	382	386	390	+18	+19
5,400	366	377	369	380	375	386	383	394	+17	+17
5,500	363	380	366	384	372	390	378	395	+15	+15
5,600	359	383	362	386	370	395	374	399	+15	+16
5,700	354	384*	362	393	369	400	373	405	+19	+21
5,800	343	379	357	394	364	402	370	408	+27	+29
5,900	–	–	351	395	358	402	364	409	–	–

The Goodwrench Quest, Part 6

By Jeff Smith
Photography by Ed Taylor

The Goodwrench 350 small-block has proven to be so much fun and so responsive to improvements that we keep coming up with more ideas to test. In the last chapter, we created a thumpin' small-block that made 408 hp and 430 lb-ft of torque, based on some carefully selected off-the-shelf parts. The key components were the iron Vortec heads, matched to a Comp Cams 268 Xtreme Energy camshaft, an Edelbrock Performer RPM intake, and a 750cfm Holley, list-number 4779, double-pumper carburetor. For ignition we used an HEI distributor and finished up with 1 5/8-inch Hooker headers and Borla 2 1/2-inch mufflers. Frankly, we were so surprised at how well the engine performed that we decided to keep going.

After a quick planning session with Chief Test Engineer Ed Taylor of Ventura Motorsports, we came up with yet another test of budget-based power. Initially we decided to run a Weiand mini-blower across the engine, but since the Weiand blower manifold didn't fit the Vortec bolt pattern, we would be hindering the blower's potential to bolt it to the stock Goodwrench heads. Instead, since our Goodwrench engine has always been targeted as a cost-conscious performer, we decided to test the new TFS aluminum 23-degree cylinder heads.

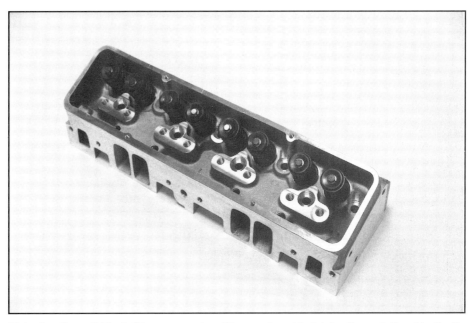

Virtually all small-block Chevys operate with an exhaust-to-intake flow relationship that's around 70 to 80 percent. The TFS heads measure a 73 percent flow at 0.400-inch valve lift. This means they can benefit from the dual-pattern camshaft. If the percentage were higher, a single-pattern cam would work better.

HEAD GAMES

The TFS 23-degree head is the latest of a new generation of cylinder heads that has surfaced out of Trick Flow Specialties (TFS). Chuck Jenckes is the engineering manager and head honcho at TFS, and the goal for this head was to create a simple, bolt-on cylinder head that offers great airflow at an affordable price.

Cylinder heads are the key to performance on any engine, and our Goodwrench 350 is a great testament to that fact. If you have kept track of the Goodwrench Quest series, we started out with stock heads, then pocket-ported them, then stepped up to a set of over-the-counter Corvette TPI aluminum heads available through GM Performance Parts. After thoroughly flogging these castings, we then stepped up to a set of iron Vortec heads, also available through GM Performance Parts. In each case, power mainly improved based on increased cylinder-head airflow.

The TFS head features a midsized 195cc intake port, which is slightly larger than stock ports that generally measure around 165 to 170cc. But size is not everything. The key is how well both the intake and exhaust ports flow since the combination of the two establishes the head's ultimate performance potential. The TFS head comes with a 65cc combustion

Small-Block Chevy Engine Buildups

PART NUMBERS

The following are part numbers for both GM and aftermarket parts used on the Goodwrench 350.

COMPONENT	SOURCE	PART NUMBER
Goodwrench 350	Scoggin-Dickey	10067353
HEI distributor	GMPP	1104067
Cyl. heads, 1.25" spring	TFS	TFS-30400001
Cyl. heads, 1.47"	TFS	TFS-30400002
Cyl. heads, 1.46" dual	TFS	TFS-3040003
Intake, Performer RPM	Edelbrock	7101
Intake gasket	Fel-Pro	1256
Head gasket, .015-inch	Fel-Pro	1094
Camshaft	Comp Cams	XE-268H-10
Roller rockers, 1.6:1	Comp Cams	1418-16
Head bolts	ARP	134-3601
Perma-Loc adjusters	ARP	300-8241
Headers, 1 5/8-in	Hooker	Application-specific
Mufflers, XR-1	Borla	40646
Carburetor, 750cfm	Holley	0-4779

Valve sizes are 2.02/1.60-inch intake and exhaust with a small 64cc combustion chamber volume. For the Goodwrench motor, this turned out to create 9:1 compression. With a zero deck height and true flat-top, four-eyebrow pistons, the compression would jump to 10.25:1.

The key to this story is the budget-based TFS aluminum heads. The intake ports are sized as a compromise between ultimate flow and good port velocity with a 195cc intake-port volume. The heads come with screw-in 3/8-inch studs, guideplates, and head bolt washers. TFS sells three different cylinder-head part numbers based on different valve springs. The heads are offered with 1.25-inch and 1.47-inch–diameter single valve springs or with a 1.46-inch–diameter, dual-spring option.

chamber size and 2.02/1.60-inch intake and exhaust valves, and comparing the intake to exhaust flow at 0.400-inch valve lift reveals an acceptable 73 percent flow relationship. This means the TFS heads would probably benefit from Comp Cams' Xtreme Energy 268 dual-pattern camshaft that offers increased duration and lift on the exhaust side.

If you compare these heads and their flow numbers, it's clear these are not the best-flowing heads on the market for their size, but they do flow better than many cylinder heads with larger intake-port volumes. What this means is that combining excellent flow with a midsized port velocity yields plenty of potential for great power at an affordable price. It would be up to our Goodwrench 350 to see how well these heads would perform.

Before we could put this rascal back on the dyno, we again went through the compression-ratio drill to ensure that static compression would not be excessive with the 64cc combustion chamber. Normally, a small chamber like this with a flat-top piston would pump the compression way up over 10:1. But the Goodwrench engine is

The Goodwrench Quest, Part 6

blessed (or cursed, if you prefer) with a cast piston well below the deck surface with a weird chamfer around the circumference of the piston. All this contributes to lowering the compression. The bottom line is the 64cc chamber works well with the Goodwrench engine's piston and combined with a thin 0.015-inch thick, rubber-coated Fel-Pro head gasket, this generates 9.0:1 compression. The best way to improve this situation would be to pull the engine apart, deck the block, and use a true flat-top piston. This would create between 10 and 10.25:1 compression. This would still work with 92-octane pump gas without detonating, and we would see a power increase. While we could easily accomplish this task, it would pull us away from the basic premise of evaluating different heads on this off-the-shelf engine. So we decided to stick with the stock short-block, ugly cast pistons and all.

The studs must be torqued in place with 45 lb-ft of torque. The best time to do this is when the heads are already bolted on the engine. This way you can adjust the guideplates to best align the pushrods with the valves.

POWER PER DOLLAR

Once Ed had reassembled the engine, he again bolted it to Ken Duttweiler's Stuska dyno and began pulling levers. The rest of the engine remained just as it had been outfitted in the previous tests, including the 1 5/8-inch Hooker heads, Borla mufflers, Edelbrock Performer RPM intake, Holley 750cfm carburetor, Comp Cams 268 Xtreme Energy camshaft, and 1.6:1 Magnum rockers. Ed set the ignition at 37 degrees of total timing and after warming up the engine, he made the first of several dyno sweeps. Despite the low com-pression, the engine still made out-standing torque, pulling over 400 lb-ft from 3,200 all the way up to 5,000 rpm. Peak torque occurred at 4,000 rpm where it cranked out an impres-sive 423 lb-ft. Moving down the chart, peak horsepower occurred at 5,700 rpm where the aluminum heads flowed enough air to crank out 416 hp.

Ed experimented with rocker ratios, timing, and jetting, but the power numbers just moved around but never improved enough to warrant the effort. The logical step is to compare the 416 hp from these aluminum TFS

Even with the small 64cc chamber, we used the thin Fel-Pro 0.015-inch stamped, rubber-coated head gasket in order to increase the compression ratio. This created an honest 9:1 compression but with a true flat-top piston and a zero deck height (versus the Goodwrench's 0.025-inch below deck height), a thicker head gasket would pump the compression to over 10:1.

Small-Block Chevy Engine Buildups

Ed torqued the heads in place using ARP head bolts. Ed starts with 40 lb-ft of torque and works up to 50 and finally 65 lb-ft of final torque.

The taller, 195cc intake port uses a Fel-Pro 1256 intake gasket to properly seal the intake port to the Edelbrock Performer RPM manifold. We did not gasket-match the intake to the head since there is little mismatch between these two components.

The best way to set the preload on hydraulic lifters is before you set the intake manifold in place. This way you can easily find zero lash. We used ARP Perma-Lok adjusters for the Comp Cams Magnum 1.6:1 roller-tipped rockers.

heads to the iron Vortec heads that made 408 hp. The iron Vortec heads are certainly less expensive at only $400 for the pair versus the Summit heads at $800 per pair. But the Vortec heads also benefited from pocket-porting. In Chapter 5, we tested the Vortec heads before and after porting and saw a 13 hp gain with no substantial increase in torque.

If you were to take a set of the iron Vortec heads to a cylinder-head specialist such as McKenzie's Cylinder Heads (which did the Vortec heads), minor pocket-porting work would cost roughly $250. Add this to the $400 initial cost, and now the TFS heads start to look more attractive, since for only slightly more money you can have the lighter aluminum heads that obviously show greater potential right out of the box. We really can't compare the unported Vortec heads to the stock TFS heads because we changed the exhaust and a couple other minor items that make this an inaccurate comparison.

Regardless of which head you choose, it's clear that just duplicating this combination guarantees you a solid 400 hp 350 that will crank out well over 400 lb-ft of torque. We plugged this TFS head power curve into a theoretical 3,500-pound Chevelle with a TH350 trans, a 2,600-stall converter, 3.55 gear, and a 9-inch–wide sticky tire 26-inches tall (like a BFG Drag Radial). Using the Racing Systems Analysis Quarter Pro program, the simulation estimates the car would run 12.20s at around 112

The Goodwrench Quest, Part 6

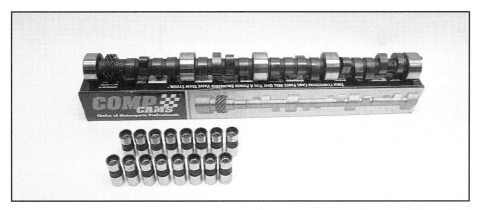

We purposely did not change the camshaft to allow us to compare this test with previous Goodwrench Quest combinations. The cam is a Comp Cams Xtreme Energy 268 flat-tappet hydraulic that has worked extremely well in several different combinations.

mph. This is plenty quick for a street car, and yet is very conservative since the engine's running through the lights at only 5,250 rpm. This is important since cast pistons don't live very long at high rpm.

The bottom line is we've come up with yet another affordable power combination with the Goodwrench 350 engine. Frankly, we're amazed this cast piston motor is still alive. We've beat on it for well over 100 dyno pulls, but we're not done yet. There's a Weiand mini-blower sitting on the shelf that we'll test in the next chapter. We plan to leave the TFS aluminum heads on the engine, but we will change to a smaller cam that may work better with the supercharger. We'll give you a hint: This baby rocks!

FLOW SPECS

TFS 195cc intake port 23-degree aluminum head, PN 30400001

Valve lift (inches)	Intake cfm	Exhaust cfm
0.050	32	28
0.100	65	70
0.200	133	102
0.300	188	140
0.400	226	164
0.500	250	183
0.600	N/A	N/A

I/E @ 0.400 lift = 73%

Flow-bench testing was performed at Westech Performance on a Superflow 600 flow bench at a test depression of 28 inches of water.

POWER CURVE

The following test reveals the entire power pull from 2,500 to 5,800 rpm. Note how the motor is making excellent torque throughout the entire run. The 350 makes at least 400 lb-ft of torque from 3,200 to 5,000 rpm with a maximum of 423 at 4,000 rpm. This would make an excellent street combination. If you wanted to increase torque at a slight loss of horsepower above 4,500 rpm, you could substitute the smaller Comp Cams 262 Xtreme Energy cam.

TEST 1

RPM	TQ	HP
2,500	357	170
2,600	358	177
2,700	361	186
2,800	375	200
2,900	379	209
3,000	391	223
3,100	395	233
3,200	403	245
3,300	407	255
3,400	412	267
3,500	415	277
3,600	419	287
3,700	421	296
3,800	422	306
3,900	422	314
4,000	423*	322
4,100	423	330
4,200	421	336
4,300	420	343
4,400	419	351
4,500	417	358
4,600	417	365
4,700	416	372
4,800	413	378
4,900	409	382
5,000	403	384
5,100	398	387
5,200	397	393
5,300	397	401
5,400	394	405
5,500	391	409
5,600	387	412
5,700	383	416*
5,800	374	413

*Denotes peak TQ and HP

Blower cams

The Goodwrench Quest, Part 7

By Mike Petralia
Photography by Ed Taylor

Lots of horsepower for little money can make you happy, but it's not always easy to find. In the case of our ever dependable Scoggin-Dickey-supplied Goodwrench Quest 350, we had come upon an impasse. In order to make more power and still retain all of its great street manners, we could either pump the 350 full of nitrous oxide and refill bottles every week or bolt on a blower and see how much power we could pump out of it.

While nitrous may be the current rage at the track, a blower offers several advantages on the street over chemically induced horsepower. For starters, the biggest advantage a blower offers is power on demand. You nail the throttle and the tires light up. That's it. No bottles to fill, no valves to open, or buttons to push—just hammer it and smile. We wanted that kind of performance for our Mouse, so we contacted the folks at Holley Performance Products and they recommended a Weiand 142ci Pro-Street supercharger. Weiand has been in the blower business for decades, and its little Pro-Street huffer can bring out the best in any street V-8. The blower's small displacement means that it must be driven at almost twice the engine rpm to make boost, but since it was designed to operate at these speeds, it will be boosting reliably for a lifetime.

Comp Cams produced a special grind hydraulic-blower cam featuring the same duration figures as our normally aspirated cam, but less lift and a wider lobe separation angle (see Cam Specs). Comp Cams would be happy to grind one for you.

BLOW TO GO

Once again, we asked master Mouse handler Ed Taylor of Ventura Motorsports to strap the Goodwrench 350 onto Ken Duttweiler's dyno for some power pulls. But before he could do that, we decided to make a few changes. First, to make a blower work really well on the street and produce all the power it's capable of, it should be combined with a blower-compatible cam. Not all cams work well with blowers, but you can make the most power and always run at peak performance once you've installed the proper grind. Typically, blower cams are listed in all the major cam grinders' catalogs and picking the correct one is as simple as calling the companies on the phone. A blower cam will have a wider lobe separation angle, which decreases the amount of overlap and helps keep the freshly boosted air/fuel charge in the cylinder.

We contacted Comp Cams for a blower cam for the Goodwrench Quest 350, which featured identical duration figures, but slightly less lift than the XE268-H cam we ran in the normally aspirated engine. Comp Cams spread the lobe centers apart by grinding the cam with a 114-degree lobe separation angle, and Taylor

Small-Block Chevy Engine Buildups

Ed Taylor of Ventura Motorsports installed the new cam straight-up. Most blown street engines don't need the cam advanced to make tons of bottom-end power, and ours was no exception. Taylor applied a locking compound to the ARP timing-chain bolts and torqued the Comp Cams chain set to 25 lb-ft.

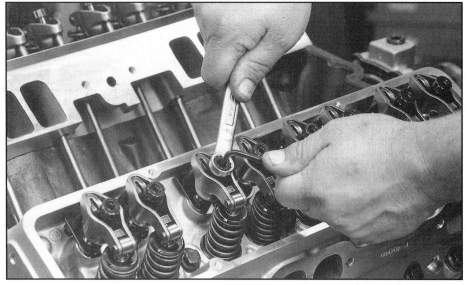

Taylor adjusted the Comp Cams 1.5:1-ratio roller-tip rockers to zero lash plus an additional 1/2 turn.

installed it straight-up.

After bolting on the TFS aluminum heads in Chapter 7, the small block calculated a 9:1 compression, which was perfect for a street blower making under 6-psi boost. Considering that the TFS heads are made from aluminum and feature an efficient combustion chamber shape, we probably could have safely increased compression to 9.5:1, but that would have meant tearing the short-block apart and milling the deck. For the minor increase in power it may have netted us, we determined that was something we didn't want to do. Although a regular GM HEI would be more than adequate under the low boost conditions you'd normally see on the street, we asked Taylor to install a Holley billet distributor with matching plug wires to ensure that the new boosted air/fuel charge would not have trouble igniting. Everything else in this engine was compatible with the blower, so it was time to visit the dyno.

PUMPING POWER

Dyno-testing a blown street engine such as this one is a pretty straightforward process. You bolt on a carb that's big enough to ensure the blower will get all the air it needs to make boost, jet it a little rich to be safe, decrease the total ignition timing a bit to stave off detonation, and then lean on it hard. That's exactly what Duttweiler did, and the Goodwrench Quest 350 running on 92-octane pump gas reported back sharply with 471 hp. Even more fun was the huge increase in low-end torque we found over its normally aspirated configuration. The dyno began recording data at 2,500 rpm, and the normally aspirated engine made a respectable 357 lb-ft of torque down there. With the blower bolted on, torque at 2,500 rpm jumped to more than 400 lb-ft, almost eclipsing the 410 mark. It gained more than 50 lb-ft of torque at a rpm where most of us are just cruising down the highway. Peak torque also responded well, increasing by 29 lb-ft, making 452 lb-ft at 4,000 rpm. But it's the low-end that drives you around, and this Mouse's tremendous low-end is what you'll feel in your rear!

Interestingly enough, after making several jet changes to the Holley 750-cfm double-pumper carb, Taylor found that the blown Mouse made its best power using the same jets as it did normally aspirated. While it's unusual for a blown engine not to need more fuel, this could be a case of the carburetor working more efficient-

The Goodwrench Quest, Part 7

The Weiand blower intake manifold is designed to fit a Fel-Pro intake gasket (PN 1256). Taylor used silicone on the end rails, which is the best way to seal aluminum intakes.

A blower intake doesn't look anything like the typical four-barrel manifolds we're used to, but its primary job is still to direct air and fuel into the cylinder heads. The low-profile Weiand intake means that this blower can fit under older musclecar hoods like the cowl-induction models on Camaros and Chevelles.

ly once the blower was installed. It might not have been pulling all the fuel and air it could through its normally aspirated venturis, and the higher venturi velocities created by the blower kept the engine running at peak efficiency. The dyno's BSFC (brake-specific fuel consumption) numbers hovered around the mid- to upper-4s, showing that our blown Mouse was fairly efficient in converting fuel into horsepower.

TIMED TO BLOW

Ignition timing is most critical in any blown motor—too much and your engine will turn into melted cheesecake; too little and it'll drive like melted cheesecake. For the first 92-octane pull, Taylor and Duttweiler called upon their years of turbo experience and set the total ignition timing at a conservative 28-degrees advance. The engine responded sluggishly and the exhaust temps were high, both signs that the engine wanted more spark advance. Taylor bumped the timing to 32 degrees, and the small-block came alive. Peak power was generated using 32 degrees of total advance. It's impor-tant to understand that while this number may have worked well on the dyno, it could be a bit high for the average street car. Your car may load the engine differently than a dyno, and our test day was relatively cool. Higher inlet air temperatures tend to make an engine more prone to detonation, and detonation is the bane of all supercharged engines. Detonation in blown engines can kill a set of pistons before you even hear it knocking, so if you install a blower on your mild street engine, you'd be wise to add forged aluminum pistons to the mix. Adding a shot of 100-octane fuel or octane boost may also be a good idea.

Any way you look at it, installing a blower like the Weiand Pro-Street 142 is money well spent. The power benefits are huge, the installation process is simple, and the rewards are immeasurable. Going blown is certainly the fastest way to get down the highway.

Small-Block Chevy Engine Buildups

This is a critical step that people often ignore. When installing any blower, the hold-down bolts are rarely torqued properly. Weiand's included instructions that emphasized torquing the blower bolts to 8 to 10 lb-ft (96 to 120 in-lbs) max. This is not a misprint, and any more torque may bind the blower, causing damage and voiding the warranty.

The Holley billet distributor went in after we primed the oiling system using a drill motor and an old distributor shaft. You can see Weiand's throttle-cable bracket bolted to the back of the blower, which will accept a stock GM cable's end, although a longer length cable may be required.

If your early Chevy's water neck exits right or straight out, you'll need to switch to a neck that exits out the left side (driver side) of the manifold. This may also require radiator modifications to relocate the hose for your new water neck.

Although it's not visible in this photo, Taylor is using a 3/4-inch box-end wrench on the idler-pulley bolt's nut to pull the idler arm down. He then slips the blower belt over the idler pulley. Weiand blower kits are available to fit most long and short water pumps and will clear most factory accessory drivebelts.

CAM SPECS

Comp Cams ground a special blower cam for us featuring duration figures identical to the XE268H-10 cam we'd previously run. Valve lift was decreased slightly, and the lobe centers were spread further apart to take advantage of the blower.

Cam grind No.	CS5443/5216 11114
Advertised duration	268 intake, 280 exhaust
Duration @ 0.050	224 intake, 230 exhaust
Valve lift	0.477-inch intake, 0.459-inch exhaust
Lobe separation	114 degrees

BLOWER POWER

Normally aspirated, our Goodwrench Quest 350 cranked out 423 lb-ft of torque and 416 hp. With the addition of the Weiand Pro-Street 142 supercharger and a new Comp Cams blower cam, we saw a huge increase in low-end torque and a nice bump in top-end horsepower.

	Normally Aspirated Power		Blown Power	
RPM	**TQ**	**HP**	**TQ**	**HP**
2,500	357	170	402	193
2,700	361	186	405	208
2,900	379	209	412	228
3,100	395	233	422	249
3,300	407	255	430	270
3,500	415	277	439	292
3,700	421	296	446	314
3,900	422	314	451	335
4,100	423	330	452	353
4,300	420	343	451	369
4,500	417	358	451	387
4,700	416	372	451	404
4,900	409	382	447	417
5,100	398	387	442	429
5,300	397	401	435	440
5,500	391	409	431	451
5,700	383	416	426	462
5,900	n/a	n/a	419	471
6,000	n/a	n/a	402	459

Peak HP: 416 @ 5,700   471 @ 5,900
Peak TQ: 423 @ 4,000   452 @ 4,000

9 Street Fighter 377

Building A Hot 377ci Small Block
By Jeff Smith
Photography by Jim Smart

There are almost as many variations on the small-block displacement theme as there are aspiring actresses in Los Angeles. The guys at Coast High Performance were in the process of assembling a cool combo 377ci small-block for a customer when our man Jim Smart happened along with a camera. If you recall our recent 377 versus 383 shootout, you already know that a 377 is the big-bore, short-stroke version small-block using a 0.030-inch oversize 4.155-inch bore and a 350ci engine crank (3.48-inch stroke) to create this interesting variation on the small-block theme. What's wrong with building a 406ci and adding in all those extra cubic inches? Nothing except that this owner wanted something different.

The idea behind a 377 is a higher-winding engine that can take advantage of the shorter stroke to reduce piston speed. With a 400's 3.75-inch stroke, the piston must travel an additional 1/2 inch for each revolution of the crankshaft (the quarter-inch stroke difference on the way up and on the way back down). This means at 6,500 rpm, a 400 small-block will push the pistons to an average speed of 4,062 feet-per-minute (fpm) while a 3.48-inch stroke crank motor (like a 377) will only run the pistons up to 3,770 fpm. An equation for average piston speed is: stroke x rpm/6 = average piston speed in fps (e.g., 3.75 x 6,500/6 = 4,062.50). This may not sound like much of a difference, but with heavy forged pistons and a cast crank, these speeds are near the upper end of the reliability scale at roughly 4,000 fpm. This is also one reason why 400s tend to not make as much top-end horsepower, since the longer stroke creates more friction and requires more horsepower

Street Fighter 377

Coast started with a good (and increasingly hard to find) 400 block and completely checked it over to ensure it wasn't cracked and still had a decent cylinder-wall thickness at 0.030- overbore. Once it was bored, honed, and cleaned, Coast checked all the bearing clearances to ensure everything would spin properly. Since ARP studs were added, this also required the mains to be align-honed. Never use studs without first align-honing the main journals.

Coast selected a Probe steel 350-inch stroke crankshaft to work in combination with the 400 block. This requires the larger 400-style main bearing size versus the smaller 350 main journal size. The rod journals end up at the 350 journal diameter and are connected to a set of Probe Track Master 6-inch rods.

Probe supplied the main and rod bearings while ARP studs provide the clamping force. Since the mains were align-honed, this also required one of Fel-Pro's special oversize two-piece rear-main seals.

Coast decided on an Isky mechanical-roller camshaft for the Street Fighter with 0.576-inch lift when used with a set of 1.5:1 roller rockers. You will also need to use a special bronze-tip fuel pump pushrod to prevent galling that can occur between the steel roller cam and the OEM fuel pump pushrod.

to spin at higher engine speeds.

Now that we've had our internal combustion engine theory lesson, let's look into the pieces Coast High Performance selected for this Street Fighter 377. Edelbrock's Performer RPM head is a great compromise cylinder head that offers a smaller intake-port volume of around 167 cc for excellent inlet velocity to help in making torque while still delivering good airflow numbers (See Chapter 17, Flow to Go). This head also delivers an excellent 73 percent exhaust-to-intake relationship, which means the exhaust port is capable of delivering decent horsepower even with a single pattern camshaft with no extra duration or lift on the exhaust lobe of the cam.

Coast combined the RPM head with Edelbrock's new Performer RPM Air Gap dual plane intake manifold that has proven to be an excellent intake. The dual plane pumps up the low and midrange power while sacrificing only small amounts of top-end power. All of this induction tuning is matched with a healthy Isky mechanical roller cam that offers 244 degrees of duration at 0.050-inch tappet lift with 0.576-inch lift using a 1.5:1 rocker ratio. Any cam with over 235 degrees of duration at 0.050-inch tappet lift is a big cam for the street and will probably deliver peak horsepower between 6,100 and 6,500 rpm. Matched with a set of 1 3/4-inch headers, this engine should be capable of around 450 hp at 6,100 to 6,300 rpm.

So let's take a look at what it takes to assemble a 377ci Coast High-Performance Street Fighter small-block Chevy. It might just give you a clue for your next healthy small-block.

Roller cams don't require moly-lube on the cam lobes, just a coating of thick assembly lube that will stay in place once the motor is assembled. Once the cam is in, the double roller drive also requires a thrust button to prevent cam walk. Keep this endplay dimension to between 0.005 and 0.010 inch.

With the cam in place, the next step is to pop the piston and rod assemblies in place. Using an ARP tapered ring compressor, it's easy to pop the pistons in, especially with 1/16-inch rings.

With the rotating assembly buttoned up, the Coast boys also bolted on the Mellings oil pump using a Canton oil pan pickup. Note that the brace supports the pickup, so it won't vibrate loose.

Next, Coast installed the ARP head bolt studs. Be sure to install these studs finger tight. Do not torque these studs into the block. Note that the shorter stud along the bottom row is undercut so that it will stretch the same amount as the longer upper-row studs. With the Fel-Pro head gasket in place, it's time for the heads.

The ARP studs offer minimal runout that makes installing the heads easy. The Edelbrock Performer RPM aluminum cylinder heads are equipped with larger 7/16-inch ARP rocker studs to take the additional rpm this engine will deliver. Make sure to coat all threads with sealant before installing the studs.

Street Fighter 377

Coast checked for proper pushrod length, then added a set of Manley pushrods to ensure that the valvetrain geometry would be correct. Keep in mind that roller cams generally use a shorter pushrod than flat tappet followers. Next came the Isky roller 1.5:1 rockers. You'll also note the larger-diameter springs that are necessary when using a high-lift roller cam. These are necessary to ensure proper valve control at 6,500 rpm or higher engine speeds.

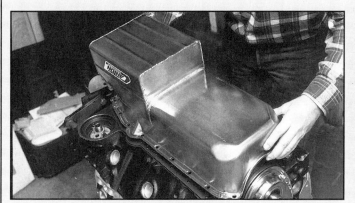 

Along with a one-piece Fel-Pro oil pan gasket, Coast bolted on the Canton aluminum oil pan. The extra sump depth adds a quart of oil while also ensuring that the oil pump pickup is always submerged in oil.

Coast test-fitted the Edelbrock Performer RPM Air Gap manifold to ensure the ports line up properly and the Fel-Pro gaskets sealed, especially at the bottom of the ports. Then, before permanently installing the manifold, the rockers were installed with the lash set.

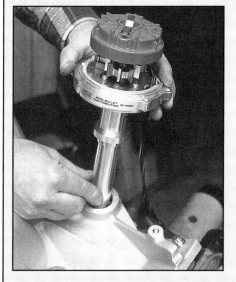

For ignition, Coast chose a Pro Billet MSD distributor and MSD-6AL box to amplify the spark. The MSD was also fitted with a bronze silicon gear to be compatible with the steel gear on the roller camshaft. Keep in mind that MSD distributors use a 0.5-inch shaft diameter that is larger than stock Chevrolet distributors, so you need to use an MSD gear. The last details included the Fluidampr balancer, Speed Demon 750-cfm carburetor, and the Edelbrock polished-aluminum valve covers. What hasn't happened yet is the vent system that will have to be added to the valve covers. Final details will include a set of MSD plug wires and a set of headers to tune the exhaust. Then this rascal is ready to fly!

CAM SPECS

The Isky mechanical roller cam used in this engine buildup is intended to create great midrange and top-end horsepower. As a daily-driver street engine, a better choice might be a cam with roughly 8 to 10 degrees less duration. Isky recommends 10 or 11:1 compression, a 750-cfm carburetor, and a 3.90 to 4.10 rear gear ratio. Because this is a steel-billet roller cam, it also requires a cam button and a bronze distributor drive gear.

Cam: 201575
Duration @ 0.050 Int/Exh: 244/244
Lift Int/Exh: 0.576/0.576
Lobe Separation Angle: 108
Lash Int/Exh: 0.028/0.028

PART NUMBERS

Component	Manufacturer	PN
Block, used 400-style	OEM	—
Crankshaft, 3.50-inch stroke	Probe	110210034
Rings, Dura-Moly, file-fit	Childs & Albert	SR4002
Pistons, 0.030-over flat	Probe	1303-10821
Connecting rods, 6-inch	Probe	SRD35060
Main bearings	Clevite	160310113
Rod bearings	Clevite	160610530
Cam bearings	Clevite	160110362
Oil pump	Melling	M-55A
Oil pan, aluminum, 6 qt.	Canton	15-040
Oil pump pickup	Canton	20-060
Timing chain cover	Canton	21-900
Camshaft	Isky	201575
Roller tappets	Isky	1241-LSH
Roller rocker arms, 1.5:1	Isky	204-716
Oil pan gasket	Fel-Pro	1802
Head gasket	Fel-Pro	1003
R.A.C.E. gasket set	Fel-Pro	2702
Rear main seal, oversize	Fel-Pro	2909
Intake gasket	Fel-Pro	1206
Valve cover	Fel-Pro	1628
Heads, Performer RPM	Edelbrock	6073
Intake manifold, RPM Air Gap	Edelbrock	7501
Valve covers	Edelbrock	4153
Water pump, long	Edelbrock	8811
Distributor	MSD	8630
Plug wires, universal	MSD	3108
Carburetor, Speed Demon 750	Demon	1402010

Agent 87, Part 1

How to Make Great Power on Cheap Gas
By Jeff Smith
Photography by Ed Taylor

10

The price of performance has just shot up with the cost of premium 92-octane fuel pushing $2.00 per gallon on the West Coast and even higher in other areas. With prices of 30 to 40 cents a gallon cheaper for 87-octane gasoline, the less expensive fuel now looks more attractive than ever. The quest for more power has always been a game of compromises between more cylinder pressure versus the higher-octane fuel that these higher cylinder pressures demand.

However, the real key to making power with low-octane fuel comes down to cylinder pressure and ignition timing. In all the dyno testing that the magazines and performance-engine builders do, everyone uses 92-octane pump gas or some form of higher-octane race gasoline. This is because high-octane fuel burns more slowly than lower-octane fuel. This slower-burning fuel allows the engine builder to increase static compression ratio to make more power.

Most enthusiasts assume that higher compression ratios add a serious amount of power to an engine. But depending on several variables, small amounts of additional compression only marginally increase power. A classic rule of thumb for compression predicts that a single point of compression ratio is worth roughly a 3 to 4 percent increase in power. Therefore, bumping compression from 9:1 to 10:1 on a 350 hp small-block, for

example, would increase peak horsepower only 10 to 15 hp. While this is certainly a good move, it's not always a given. We've tested engines where a 1/2-point increase in compression was worth no additional power.

While conventional wisdom holds that static compression should remain around 9:1 for most street engines, that number can change based on the selection of other components such as iron or aluminum cylinder heads, and especially with respect to camshaft profile. Large camshafts tend to bleed off cylinder pressure at low- and mid-rpm engine speeds because of the effect of overlap, so these camshafts generally demand more static compression ratio compared to an engine with a shorter-duration camshaft with less overlap.

For simplicity's sake, we decided to concentrate on something basic, like the relationship between ignition timing and torque. Many enthusiasts believe that max power requires advancing the timing until the engine detonates (rattles) at wide open throttle and then retarding the timing a little. While this is OK for a quickie blast down the street, an ignition-timing total of around 36 degrees before top dead center (BTDC) is usually much closer.

Advancing ignition timing means the combustion process starts sooner. If you were paying attention in high school science class, you were probably taught that combustion is like an explosion inside the cylinder. While this is true, combustion does

Small-Block Chevy Engine Buildups

This may not look like an impressive small-block, but how many Mouse motors do you know that can make 411 hp on 87-octane gasoline? This 355 is very similar to our original Goodwrench Quest engine with Trick Flow heads, 9:1 compression, and a Comp Cams 268 cam. The main difference is this motor made this power on cheap gas.

One key to this engine is the slightly dished Speed-Pro forged pistons that, when combined with the 0.005-inch in-the-hole deck height and Fel-Pro 1003 head gasket, create a 9.4:1 compression ratio. The Hye Tech short-block also features a CAT 5140 forged steel crank and rod package.

take time. If you light the spark plug too soon (44 degrees BTDC versus 36 degrees for example), sufficient pressure builds in the cylinder before the piston reaches TDC. If that happens, the other seven cylinders must overcome the pressure and the engine loses power. Therefore, for any given engine com-bination, there is an ideal ignition-timing figure that will produce maximum power.

In all of CHP's dyno tests, we run through a number of timing tests to determine the optimal ignition timing. For this story, we used a 355ci small-block short-block assembly from Hye Tech Machine using 9.4:1 compression, a CAT forged 5140 steel crank and stock rod package, a set of Speed-Pro's latest-design dished pistons, and a Mellings oil pump. For a camshaft, we chose a Comp Cams Xtreme Energy 268 cam to bump the valves in a pair of Trick Flow aluminum 23-degree 2.02/1.60-inch valve heads. To this we added an Edelbrock Performer RPM intake, a Holley 750-cfm double pumper carburetor, and a set of Hedman 1 5/8-inch Elite headers. If you've been reading CHP for the last year or so, then you may recognize many similarities between this engine and the Goodwrench Quest engine that we tested last year. Think of this as the Son of Goodwrench Quest.

THE TEST

The plan was to check ignition timing on the dyno starting at 30 degrees of total timing and then bumping the timing up two degrees through 38 degrees total using cheap 87-octane fuel to see when it would detonate. The only problem was the engine never detonated! While we limited the testing to a maximum of 38 degrees of total lead, this 400 hp-plus small-block never rattled on 87-octane fuel!

Much of this can be attributed to two important points. First, the inlet air temperature was a very balmy 70 degrees while the engine temperature hovered around a reasonable 180–190 degrees. Both of these conditions contribute greatly to reducing the chance for detonation. This is because cooler air in the engine reduces the tendency for an engine to detonate.

If the engine is also at a cooler temperature, this means there is less heat for the inlet air to pick up off the intake manifold and port walls on its way to the combustion chamber.

We know that the engine didn't detonate, since we were using a sophisticated J&S Electronics detonation sensor and control system that Duttweiler uses when testing his high-output, turbocharged engines. This system uses a GM knock sensor that picks up detonation frequencies transmitted through the block and signals the J&S-designed microprocessor that this has occurred. The processor calculates which cylinder was firing to create that detonation and then pulls the timing back in the cylinder that's actually rattling. This allows the more sophisticated engine

Agent 87, Part 1

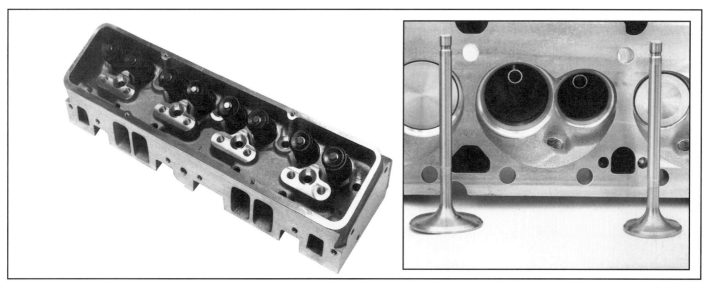

We chose the aluminum Trick Flow 23-degree cylinder heads for several reasons. First, the aluminum tends to transfer heat from the chamber much quicker than iron heads, which is one reason why this engine didn't detonate on 87-octane fuel. The general rule is to reduce compression by roughly one full point when using iron heads versus aluminum. The second reason is the well-designed, 64cc chamber that requires less total timing to create maximum power. Those are 2.02/1.60-inch stainless valves at right.

Hye Tech supplied the short-block using four-bolt main caps, a CAT steel crankshaft and stock rods. ARP supplied the main studs along with Speed-Pro rings and bearings.

tuner to extract the maximum power from his engine while minimizing the damage risk of detonation.

THE RESULTS

As you can see from the dyno testing, it is certainly worth the effort to experiment with ignition timing. It's obvious that at 30 degrees of total lead, the engine is struggling. Compared to 38 degrees of total lead, there is as much as 22 lb-ft of torque difference.

It's important to look at the entire curve of all five tests since the power peaks appear quite similar, but there are some subtle differences worth noting. For example, even though 34, 36, and 38 degrees of timing all offer 432 to 433 lb-ft values, Test 5 shows 12 lb-ft more torque at 3,400 rpm and 18 lb-ft at 2,600 rpm. These are dramatic numbers that will show up as much quicker acceleration in the car. We ran these power numbers through the Racing System Analysis Quarter Pro computer program using a 3,500-pound Camaro with the two power curves from 30 degrees of lead and 38 degrees. The difference was approximately 0.30 second and 3.10 mph between the two timing curves.

What we really wanted to show, but didn't occur with this test, was what happens to power when the engine experiences detonation. Detonation is uncontrolled combustion. When this occurs, power drops off radically. Often too, detonation can begin long before it becomes audibly apparent. Plus, this is just the horsepower price. There's also the risk of serious engine damage that occurs as a result of this uncontrolled combustion. The most common form of engine carnage is damaged or broken pistons (especially if they are cast pistons), pinched ring lands, valve damage, hammered rod bearings, and many other related

Small-Block Chevy Engine Buildups

We had such good luck with the Edelbrock RPM intake on the Goodwrench engine, we stuck with Edelbrock on this engine using instead the RPM Air Gap manifold along with a Holley 4779 750-cfm double-pumper carburetor.

The cam timing choice for this 87-octane test went to the Comp Cams 268 Xtreme Energy cam because it has worked so well in many of our past engine buildups. The combination of just enough overlap with the dynamic compression allowed this engine to make exceptional power on 87-octane fuel. Specs are: Duration @ 0.050 Int/Exh: 224/230; Lift Int/Exh w/1.5:1 Rockers: 0.477/0.480; Lobe Separation Angle: 110

We also used Fel-Pro gaskets throughout to seal up this small-block. Here we used a Fel-Pro intake gasket on the Trick Flow aluminum heads.

engine maladies.

This test does illustrate the result of having to pull timing back in order to control detonation. Let's say that 36 degrees of timing is what your engine really needs, but unfortunately, it detonates like eight rocks in a Folger's coffee can at wide open throttle. Let's say it takes pulling 6 degrees of timing out of the total advance curve to eliminate the detonation. If this theoretical example were anything like our dyno test engine, retarding the timing from 36 degrees total to 30 degrees would cost an average of 17 lb-ft of torque across the entire power band of this engine. That's some serious power that you've had to give up to prevent detonation.

CONCLUSION

Is it possible to build an engine that will make over 400 hp on 87-octane fuel? The obvious answer is a resounding "Yes!" but with a few warnings. First, it's imperative that you pay close attention to the timing curve as well as cool inlet air temperature, cool engine temperature, and employing a well-designed combustion chamber that doesn't require as much timing to make the same power. These concepts, along with a few dozen others, are all a part of making power in today's performance world. If you can make equal power without having to spend that extra 30 cents a gallon for 92-octane fuel, that money could be saved to buy that next nitrous kit or perhaps that trick six-speed trans you've been lusting after. How would you like to spend your money?

We used an MSD billet distributor along with an MSD-6A spark box to fire the mixture, while a clip-on sensor senses when No. 1 cylinder fires. When tied into the ignition system, the J&S Safeguard system (bottom photo) senses detonation and retards the timing sufficiently to prevent detonation. This is the J&S professional model that Duttweiler uses, but J&S also makes a more affordable street system as well.

Agent 87, Part 1

POWER ON DEMAND

RPM	TEST 1 30 Deg. TQ	HP	TEST 2 32 Deg. TQ	HP	TEST 3 34 Deg. TQ	HP	TEST 4 36 Deg. TQ	HP	TEST 5 38 Deg. TQ	HP	MAX DIFF TQ
2,600	358	177	363	180	376	186	378	187	381	189	23
2,800	373	199	382	203	388	207	390	208	391	209	18
3,000	385	220	389	222	399	228	401	229	402	230	17
3,200	394	240	400	244	407	248	408	248	410	250	16
3,400	400	259	407	261	415	269	416	269	419	271	19
3,600	413	283	418	286	426	292	427	292	429	294	16
3,800	420	304	427*	309	431	312	431	311	432	312	12
4,000	419	319	427	325	432*	329	432*	329	433*	330	14
4,200	420*	336	427	341	431	345	431	345	432	345	12
4,400	418	350	425	356	430	360	430	361	432	361	14
4,600	412	360	421	369	428	375	428	375	431	376	19
4,800	403	369	411	375	419	383	420	384	421	385	18
5,000	390	371	400	381	410	390	409	390	414	394	24
5,200	385	381	395	390	404	400	403	399	406	402	21
5,400	378	388	388	399	396	407	397	408	398	409	20
5,600	364	388*	375	400*	383	409*	384	411*	386	411*	22
5,800	334	369	345	381	355	393	357	394	361	398	27
6,000	315	360	334	381	343	393	345	394	355	406	40
Avg.	388		396		404		404		407		19

*Indicates peak TQ and HP levels

The MAX DIFF column on the far right shows the maximum torque change between 30 and 38 degrees of timing. This represents as much as 40 lb-ft of torque at 6,000 rpm, which is an amazing 46 hp.

11 Agent 87, Part 2

390 HP on a Budget 355 cid
By Jeff Smith
Photography by Ed Taylor

In the last chapter, we coaxed 411 horsepower out of a 350 small-block on pump gas. Obviously, the small block used high-quality parts to make this kind of power. Hye Tech Performance built the engine using a CAT steel 5140 crank, good forged Speed-Pro pistons, and aluminum TFS 23-degree cylinder heads. This engine responded with great power on 87-octane fuel especially with its 9.4:1 compression ratio. While this was a great example of how to make good power, we felt we could make similar power with a much less expensive engine combination. What began as a notes-on-a-stained-napkin idea evolved into the Miserly Mouse.

The Miserly Mouse is very similar to the small-block described above, but with a few subtle differences. Based on the success of that combination, we looked at where we could save some money and yet still produce similar power. The key to this exercise comes down to matching cylinder heads and cam timing. To make decent power, you have to have a good set of cylinder heads. One half of the Miserly Mouse's one-two punch is a set of the GM Performance Parts iron Vortec cylinder heads. Chapters 2–8 showed us that these inexpensive production iron heads can make great power at a very reasonable price of around $450 per pair complete and ready to bolt on.

The other half of this equation was a bit more challenging. The concept of building a performance engine that makes decent power on 87-octane fuel requires a special camshaft. The basis for this 87-octane effort revolves around a slightly lower 8.75:1 compression ratio (because of the iron heads) along with an emphasis on improving combustion efficiency.

Remember that chamber design and pistons can combine with a tight quench area to create an "active" combustion space that can help reduce the tendency toward detonation. Since detonation kills power and can destroy an engine, we want to avoid it at all costs. But savvy engine builders will tell you that you can get away with higher compression if you build the engine correctly. This is why we chose the Vortec iron heads. Not only do the heads have an excellent reputation for airflow right out of the box, but the Vortec heads also enjoy the benefit of an outstanding combustion chamber design. This heart-shaped chamber creates turbulence in the combustion space to prevent detonation.

When we conspired with Ed Taylor to build this engine, we had three goals in mind. The first was to be conservative with a compression ratio under 9:1 to suppress detonation. The second was to optimize combustion efficiency by minimizing the piston-to-head clearance. This minimal

Agent 87, Part 2

We started with a fresh two-bolt main block that Jim Grubbs Motorsports bored, honed, and align-honed for the ARP main studs. Taylor also offset the rear-main seal to prevent leaks.

The Federal-Mogul pistons come with a D-shaped dish to optimize combustion efficiency. Taylor then hand-sanded all the sharp edges to prevent hot spots that could create pre-ignition or detonation.

The only right way to assemble rod bolts is to use a rod bolt stretch gauge. This ensures that the bolt is properly tensioned to prevent over- or undertorquing. We lost an engine to undertorqued rod bolts so now we religiously use the ARP rod-bolt stretch gauge.

ARP also supplied an oil pump stud to mount the Mellings oil pump and an oil-pump drive.

clearance enhances turbulence in the chamber that should suppress detonation. The third was to use a dished piston, but to choose a D-shaped cup instead of a full dish. The flat portion of the piston matches the flat portion of the combustion chamber to create a quench, or squish area between the chamber and the piston that creates turbulence.

Savvy engine builders know that minimizing the piston-to-head clearance in this quench area to less than 0.040-inch dramatically increases the amount of turbulence created in the combustion space between the piston and the chamber. Increasing this activity minimizes the engine's sensitivity to detonation. As a result, Taylor chose Speed-Pro forged pistons with a D-shaped 21cc dish. Add this to the Vortec head's 64cc chamber and we end up with the equivalent of an 85cc combustion chamber size with a flat-top piston.

Then Taylor mocked up the engine with all eight pistons, measured the piston deck height, and then milled the block to create as close to a zero deck height as possible. He then added a Fel-Pro composition head gasket with a compressed thickness of 0.041 inch that created an average piston-to-head clearance for all eight pistons of 0.044 inch. In discussions with race engine builders, it is possible to build a street engine with a slightly tighter piston-to-head clearance, but this requires a very tight piston-to-wall clearance to prevent piston rock at TDC. We decided to stick with our 0.044-inch clearance.

THE ENGINE

Since we intended to build a budget-based engine, we could have used a stock Chevy crank, but for just a little more money, we picked up a cast Scat crank to match our resized stock rods. Jim Grubbs Motorsports (JGM) performed the machine work on our stock block including a 0.030-inch bore, torque-plate honing, connecting rod resize with ARP bolts, and an align-hone to accommodate the ARP studs for the two-bolt main caps. You should always align-hone the mains anytime studs are used

57

Small-Block Chevy Engine Buildups

PARTS LIST

Component	Manufacturer	PN
Crankshaft	Scat	10442
Piston, dished	Speed-Pro	2441-030
Rings, 5/64-in	Speed-Pro	R9343-35
Bearings, main	Federal-Mogul	139M1
Bearings, rod	Federal-Mogul	CH7100
Bearings, cam	Federal-Mogul	1235M
Oil pump	Mellings	M55HV
Head gasket	Fel-Pro	1003
Intake gasket	Fel-Pro	1255
Oil pan gasket	Fel-Pro	1821
Timing set gasket	Fel-Pro	2702
Oil pump drive	ARP	135-7901
Oil pump stud	ARP	230-7001
Main studs	ARP	134-5401
Head bolts	ARP	134-3601
Damper bolt	ARP	14-2501
Perma-Loc	ARP	300-8241
Dress kit	ARP	534-9801
Intake bolts,	ARP	134-2002
Vortec heads	GMPP	12558060
Camshaft, Xtreme	Comp Cams	XE-268
Rocker arms, 1.5	Comp Cams	1417-16
Rocker arms, 1.6	Comp Cams	1418-16
Timing chain	Comp Cams	2100
Timing cover, 2 pc	Comp Cams	210
Pushrods	Comp Cams	760-16
Intake manifold	Edelbrock	7116
Carburetor, 625	Demon	n/a
Distributor, HEI	Pertronix	D1001
Plug wires, 8 mm	Pertronix	808290
Headers, Elite	Hedman	68298
Wire separators	Made For You	50-956
Water pump	TD Perf	8967
Valve covers	TD Perf	9702
Bolt kit, covers	TD Perf	9812
Timing cover	TD Perf	4934
Breathers	TD Perf	4871
Oil pan	TD Perf	9726
Dip stick	TD Perf	4957
Water neck	TD Perf	9229
Damper	TD Perf	8940

since the studs change the load distribution on the main caps that will affect main bearing clearance. Speed-Pro also supplied the rod, main, and cam bearings along with the 5/64-inch ring package.

Once JGM performed the final deck operation on the block to establish the proper deck height, Taylor spent a few hours assembling the short-block. This led us to the camshaft. Since we have had excellent luck with the Xtreme Energy 268 camshaft from Comp Cams, we didn't have to burn the midnight oil to come up with that decision. The cam specs are listed in the Cam Specs sidebar, but with the later closing intake valve, we figured the engine will not see detonation, even with 36 to 38 de-grees of total ignition timing.

The final assembly included a Mellings oil pump, a Hamburger oil pan from TD Performance, and a two-piece timing chain cover from Comp Cams that will allow us to make cam timing changes if necessary. TD Performance also supplied a set of centerbolt valve covers, an aluminum water pump, and an SFI-approved harmonic balancer. Pertronix supplied an HEI distributor and plug wires, while Made For You supplied the wire looms. Finally, Taylor bolted on the Edelbrock Vortec Performer RPM intake and a Road Demon carburetor to handle the induction duties while a set of Hedman Elite 1 5/8-inch headers supplied the exhaust ducting.

THE TEST

Ken Duttweiler once again offered his dyno cell to flog our Agent 87 engine. After warming up the engine and performing a short break-in period, Taylor pulled the handle on Duttweiler's dyno and started with the

Agent 87, Part 2

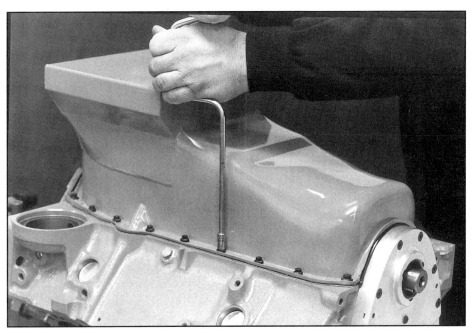

We also wanted to try one of the new Hamburger oil pans from TD Performance, sealed up with a complete set of Fel-Pro gaskets.

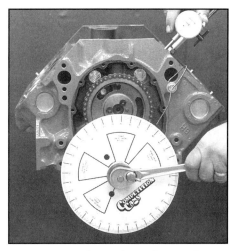

To ensure the Comp Cams Xtreme Energy cam is in the right place, Taylor degreed the cam and found the cam to be exactly where it was supposed to be at its 108-degree intake centerline.

After the block was decked, Taylor cleaned the block and all the rotating assembly and then carefully installed the rings on the pistons and pressed them into place using an ARP tapered ring compressor. Taylor also opened up the second ring end gap to 0.024-inch, with the top ring gap at a tighter 0.018-inch.

tuning process. Taylor tried several jetting configurations on the Road Demon 625-cfm carburetor but the power seemed best at the stock jetting numbers. We also started at 36 degrees of total timing, and after several pulls also discovered that the engine preferred 36 degrees.

After several pulls, Taylor went back to the stock timing and jetting combination and made several pulls to confirm the power numbers. The combination of the Xtreme Energy cam, Vortec heads, and a tight quench resulted in a great power curve that we pulled clear down to 2,400 rpm. The 355 managed over 400 lb-ft of torque at 3,300 with a peak torque of a stout 427 lb-ft at 4,000 and then delivered 390 hp at 6,000 rpm. This is an outstanding 2,000-rpm powerband between peak torque and peak horsepower. The engine also was still pulling 375 hp at 6,300 rpm. What this means is that the engine offers a wide powerband that has the potential to run low 13s in a 3,600 pound Chevy with easy regularity. Remember now, this is on 87-octane pump gas.

We also did a little quick money-saving math. If the price differential between 87 octane and high 92 octane is 25 cents per gallon and you use an average of 20 gallons of gas per week, that's a savings of $260 per year just on fuel. That's not a ton of money, but it's enough to allow you to save enough money in gas to pay for a nitrous system! So if you're thinking about building a mild street engine, this might be the way to get your horsepower and save money too. It doesn't get any better than that!

Small-Block Chevy Engine Buildups

We have had excellent luck with the Edelbrock Vortec Performer RPM intake, but the Vortec heads do require special bolts as well as a special intake gasket.

Since we were on a budget, we opted for a smaller, 625-cfm Road Demon carburetor with vacuum secondaries. Despite its small size, the carb delivered an excellent fuel curve and didn't hurt the overall power. We tried a larger, 750-cfm carburetor but saw no increase in power.

TD Performance supplied all of the smaller but necessary parts including the balancer, valve covers, thermostat housing, dip stick, and a chrome timing-chain cover.

DYNO TEST

Rpm	TQ	HP
2400	351	152
2600	376	174
2800	384	201
3000	384	220
3200	394	240
3400	408	264
3600	419	287
3800	425	307
4000	426	325
4200	425	340
4400	421	352
4600	411	360
4800	404	370
5000	396	377
5200	388	384
5400	372	382
5600	359	383
5800	352	389
6000	341	390
6200	326	385

Agent 87, Part 3

By Jeff Smith
Photography by Jeff Smith

There are no excuses when it comes to taking a car to the dragstrip to see what it will do. Since we race cars and not dynos, we decided to stuff our Agent 87 motor into our pal Pat Peterson's orange '65 El Camino, which was just sitting in his backyard waiting for an engine. So Pat and his dad Jim dropped the Agent 87 355 in along with a B&M TH-350 automatic and a 2,000-stall, 11-inch converter. Pat's El Camino also had a 12-bolt with a 4.10 gear out back that was guaranteed to spin the tires.

The plan was to bolt this Agent 87 engine in the car with as many of the same components as we used on the dyno. The Hedman Elite 1 5/8-inch headers went in with the engine, and we bolted the motor up to the B&M converter with a B&M flexplate. We also used a set of the new polyurethane engine and trans mounts from Energy Suspension to keep our rompin' small-block from pulling apart the stock rubber mounts. To keep the weight down and provide excellent cooling, we also installed an aluminum Be Cool radiator along with an excellent Spal electric fan using a slick set of Be Cool mounts. Since this is a street car, we needed a performance exhaust as well. Flowmaster makes a great Chevelle/El Camino exhaust system kit that consists of 2 1/2-inch lead-down pipes that allow sufficient room for a set of 50-series mufflers as well.

With the engine in the car and the exhaust properly plumbed, it only took a little extra work to dial in the carb and ignition. Everything on the engine remained the same, including the Demon 625-cfm carburetor, Edelbrock Performer RPM Vortec intake manifold, and the Pertronix HEI distributor. To get the car running, we set the initial timing at 12 degrees. The Pertronix distributor comes with a rather conservative mechanical-advance curve that is not all in until well past 4,500 rpm, so we swapped in a set of lighter springs to bring the curve in a little sooner. Combined with the initial timing, this gave us 36 degrees of total timing. With the vacuum advance connected, we had well over 40 degrees of timing at part-throttle, which should help the fuel mileage. We then set the idle mixture using a vacuum gauge and managed to generate around 14 inches of manifold vacuum with the trans in neutral at 900 rpm. Finally, we dropped in a K&N filter along with one of K&N's new Xtreme filter lids to give the engine every advantage to breath deeply. Since engine builder Ed Taylor had test-jetted the engine on the dyno and found the out-of-the-box jetting worked best, we left that as it was.

A quick test-cruise around the block revealed that the motor was very responsive to throttle with no hiccups or hesitations. The big question was

Small-Block Chevy Engine Buildups

Agent 87 bolted right into Pat Petersen's orange '65 El Camino with a minimum of hassle. We made a concerted effort to retain as many of the dyno-test components in the car as possible so that power would be the same.

We used the exact same set of Hedman Elite 1 5/8-inch headers from the dyno test in the El Camino. Surprisingly, we had to dimple the right-side headers slightly to clear the frame. Usually, the Hedman headers bolt right in with no problems.

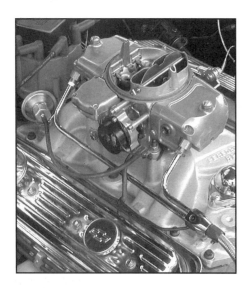

The induction system remained exactly the same as on the dyno test including the Edelbrock Performer RPM Vortec intake and the 625-cfm Road Demon carburetor. For this initial test, we left the stock jetting in the carburetor. We also employed a K&N filter with the high-flow Xtreme lid.

This '65 El Camino is Pat Peterson's first street machine. Since it's not his daily driver, he replaced the marginal 3.08 10-bolt with a 4.10-geared 12-bolt. Combined with the B&M 350 trans and Holeshot converter, the drivetrain is one reason why this El Camino is so quick. RSA's Quarter Pro simulator predicts the 4.10 gears are roughly 0.35-second quicker than a set of 3.30s.

whether this combination would detonate on 87-octane fuel in normal street driving. The weather also played a part in our initial testing. With unseasonably chilly weather in Southern California in February, cooler air prevented any kind of true hot-air detonation test since hotter inlet air temperatures increase the tendency for an engine to detonate. We did have a couple of 80-degree days, and Patrick reported that no matter how he loaded the engine, it never rattled. A couple of days of driving doesn't prove the point, but it did indicate that detonation shouldn't be a serious problem. Keep in mind that this was with 12 degrees of initial timing and a relatively fast ignition curve. Another point worth mention-

Agent 87, Part 3

Flip's Tire mounted and balanced our Mickey Thompson E.T. Street tires on a set of Center Line 15x8 Triad aluminum wheels with 5 1/2-inch backspacing. These are excellent tires that eliminate any trace of tire spin. Just bolting on these tires was worth 0.3 seconds versus the normal street tires.

We also added a Be Cool radiator and Spal electric fan to the El Camino to ensure it would run cool enough to keep detonation at bay.

ing is that the El Camino was equipped with a 4.10 gear and a loose 2,000-stall converter. A four-speed car with taller gears would generate greater off-idle loads that could induce some low-speed detonation that's not present with the automatic. However, cutting back on initial timing would probably eliminate that problem.

After 150 miles of street driving, it was time to load up and head to the track to see just what this rascal could do. Prior to the trip to the track, we plugged the horsepower curve into the Racing Systems Analysis Quarter Pro simulator. This program assumes a perfect chassis and estimated the El Camino would run in the low 13s (uncorrected for the 3,000-foot altitude) at Los Angeles County Raceway's (LACR). Since this was a test of how well the engine works in the car, we intended to run the El Camino with a set of sticky DOT-legal street tires to eliminate tire spin. Unfortunately, we overlooked the fact that Pat's El Camino was fitted with a wider, second-generation Chevelle 12-bolt that pushed our initial 5-inch backspacing tire and wheel package into the outer fenderwells. This forced us to use Pat's street tires that consisted of a set of 15x8 Weld Wheels with a 5 1/2-inch backspacing mounted with a set of ancient Mickey Thompson S/S tires in the rear.

We performed our first test on a Friday grudge night and the temperature was in the mid-40s. While this was good for the engine, traction was at a minimum. Despite severe tire spin problems at the launch and on each gear change (including Third gear), the El Camino ran an uncorrected 13.88 at 99 mph. This corrects to a 13.43 at 102.26 mph at sea level. The correction takes into account the thin air present at 3,000 feet above sea level but does not compensate for air temperature or atmospheric conditions. LACR's 60-foot clocks were not working, but we estimated the 60-foots at no quicker than 2.20s because of the cold track conditions.

The following week, we lined up a pair of Center Line 15x8 Triad wheels and mounted a set of 26x8 Mickey Thompson DOT-legal E.T. Street tires that we knew could handle all the torque our Agent 87 motor could crank out. We also decided to loosen up the front and rear suspension by adjusting the Competition Engineering shocks to full soft on all four corners. Pat had already removed the front sway bar. The next trip to the

The second trip to LACR was much more fun with the El Camino hooking up with the better ET Street tires mounted on a set of Center Line 15x8 Triad wheels.

Small-Block Chevy Engine Buildups

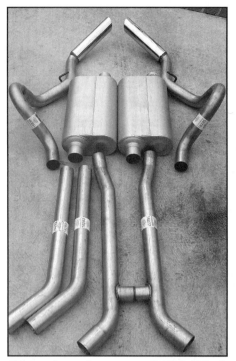

Flowmaster makes an excellent 2 1/2-inch exhaust system for Chevelles and El Caminos. The El Camino's longer wheelbase allows plenty of room for the largest Big Block II mufflers from Flowmaster, but Pat opted for the 50-Series mufflers that are louder—much to Pat's liking.

We also used the new K&N Xtreme lid to match the K&N air filter. Eventually we'd like to build a cold air box to mate to the fiberglass cowl-induction hood on Pat's El Camino.

dragstrip was much warmer with the temperature in the 60s. With the better traction afforded by the M/T tires, the El Camino immediately responded with a best time of 13.571 at 99.51-mph pass that corrects to a sea level 13.13 at 103.04-mph pass. After a dozen passes, this was the best, but the El Camino ran three other passes in the low 13.20s to back up this run.

At this point we ran out of time but since we're this close, we're going to really push this 3,700-pound cruiser until it runs at least a 12.99. Even if the El Camino doesn't knock down the 12-second door, low 13s at 103 mph is more than respectable from this Agent 87 engine.

But we knew there was more power to be made with 87-octane fuel, so we went home, put our tuner caps on, and decided to look a little closer at how the car ran. It was obvious from watching the El Camino on the starting line that it was lazy despite its 2,000-stall B&M torque converter and 4.10 gears. It was time for more torque.

Experience told us that we needed more timing at low engine speeds. This additional timing below 3,000 rpm would help build additional cylinder pressure lost from the combination of a relatively big cam and low static-compression ratio. This would make more torque and improve the e.t.—at least that was the theory.

Our original curve offered 12 degrees of initial timing with 36

The key to running faster is the careful integration of all components into a package that works best. Rather than throw more expensive parts at this project, we decided to tune what we had. After determining that the ignition curve was too slow, Tim Moore pulled the distributor for some custom work.

Agent 87, Part 3

We experimented with lightweight springs that brought the curve in slightly sooner, but this still didn't deliver enough timing at low speeds. The lightweight springs also tended to bounce the timing around at idle.

All of our tests were performed with a standard MSD timing light and an MSD tape on the balancer. You can use the engine like a distributor machine by having a friend rev the engine while you read the advance. Make sure to remove the vacuum-advance hose from the distributor before you check the timing. Also make sure the vacuum-advance hose is connected to the ported vacuum connection on the carburetor.

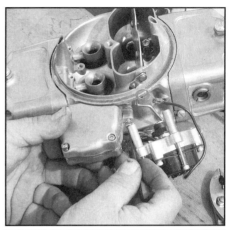

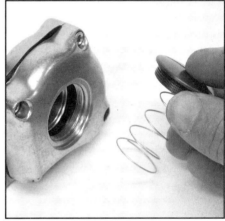

We also messed with the vacuum-secondary springs on the carburetor to open the secondaries quicker than the stock spring. Unfortunately, this didn't help the e.t. or speed.

degrees total, giving 24 degrees of mechanical advance in the distributor. At first, we tried the lightest advance springs that Pertronix offers, but that really didn't bring the curve in any quicker. This also made the initial timing at idle very unstable. We also experimented with different advance weights with disappointing results. We knew we had to take the distributor apart (see the "Distributor Blueprinting" sidebar) to limit the total advance.

The problem was that the distributor offered 24 degrees of mechanical advance. This only allowed 12 degrees of initial timing to produce our desired total of 36 degrees. To remedy this, we modified the distributor to limit its total advance. Most distributors use a pin and slot arrangement to determine the total amount of mechanical advance that is located under the mechanical advance mechanism on stock Chevy distributors. With the distributor apart, our pal Brett Benson welded up roughly one-third of the slot. We then ground the weld flat and polished the surface with a Standard Abrasives unitized wheel and reassembled the distributor.

Once the distributor was back in the engine, we found that we had limited the mechanical advance to a mere 14 degrees, requiring 22 degrees of initial timing to achieve 36 degrees of total advance (14 + 22 = 36 degrees). Checking the curve, our 36 degrees of total was now all in by 2,600 rpm using one light- and one medium-strength spring in the advance mechanism. The original curve required over 4,000 rpm to achieve maximum advance.

Clearly we had made a radical change in the curve at low rpm. For example, the original curve offered barely 25 degrees of advance at 3,000 rpm, while our modified curve now had 36 degrees of timing at 2,500 rpm. A quick test spin revealed the El Camino felt stronger between 1,500 and 3,000 rpm. We also performed a hot start test to see if all this initial timing would make our Agent 87 engine hard to start. Ironically, the engine now starts quicker than it did before and is much more throttle responsive. If necessary, we could grind part of the slot back open again, which would give us 3 or 4 degrees more mechanical if necessary. We also were afraid that with this aggressive timing curve, the engine might detonate at part throttle, but this has not been the case. That may change, however, when the dog days of summer arrive.

Ignition curves were not the only things we tried. The 625-cfm Road Demon carburetor is a vacuum-advance carb that uses a spring in the secondary diaphragm to determine when the secondary barrels open. The stiffer the spring, the longer it takes

Small-Block Chevy Engine Buildups

We tested fuel pressure using this Auto Meter fuel-pressure gauge and discovered we only had 3 psi at idle.

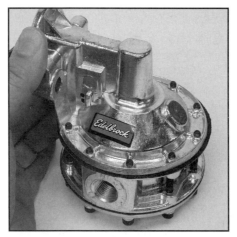

We added a high-performance Edelbrock mechanical fuel pump and now we have plenty of fuel.

Our final trick was to ice down the intake manifold and go for an all-out shot at the clocks. This is an old Stock Eliminator trick usually worth a few hundredths of a second.

SCORECARD

Here's what we've achieved so far with our first Agent 87 engine. All e.t.'s and speeds are corrected to sea level.

Run	e.t./speed
Baseline	13.43/102.26
M/T E.T. Street tires	13.13/103.04
Better ignition curve	13.28/103.48
Added fuel pump	13.07/102.1

for the secondaries to open. Changing the spring in the stock vacuum-secondary pod is rather cumbersome, so we ordered a quick-change vacuum secondary pod that would make spring changes easier at the track.

We didn't want to make any more changes to the car until we could take it back to LACR to test, since making too many changes at once can get confusing and we wouldn't know which change had made the most improvement. We again took the El Camino up to LACR and mounted the Mickey Thompson E.T. Street tires and Center Line Triad wheels on the car to eliminate the tire-spin problem with the tires set at 18 psi.

On the first pass with the better ignition curve, the El Camino left noticeably harder, pulling a better 2.08-second 60-foot time, but we did not see any improvement on our original 13.13/103-mph benchmark. It seemed that now our problem was traction. Even with the M/T E.T. Street tires, the El Camino still spun the tires.

Pat pulled the El Camino back into the pits and we decided to try a fairly light spring to see if quicker-opening secondaries would net a lower e.t. The Road Demon comes with a fairly light spring right out of the box and the effort netted us no gain.

By this time, we had run out of track time. Once back home, we plugged an Auto Meter fuel pressure gauge into the fuel line and were greeted with barely 3 psi (normal is 5 1/2–6 1/2 psi), even at idle. We figured we had found the problem. We also decided to add an Air Lift air bag to the right side in an effort to preload the chassis and prevent tire spin on the right rear tire.

Back at the track, the first run was disappointing. Rechecking all our tuning points, Patrick drove the El Camino to a corrected 13.07/102.11, which was a small e.t. improvement. Now we only needed 0.08 second and we'd be in the 12s. Unfortunately, we had time for only one last run at LACR's Friday night Grudge Night. As a last resort, we cooled the engine down, put ice on the intake manifold, and made sure the electric fan was off before making the last run of the day. The cooler manifold had paid off before with a slight e.t. improvement, but this time it was no dice. The El Camino ran slower (there was about a 10 mph head wind that night), and we were forced to head home without our 12-second timeslip trophy. Time to go back to the drawing board.

Agent 87, Part 3

DISTRIBUTOR BLUEPRINTING

Disassembling a Chevy distributor is easy. We'll run you through the steps and also show you where to weld the mechanical-advance slot and how much to close it up.

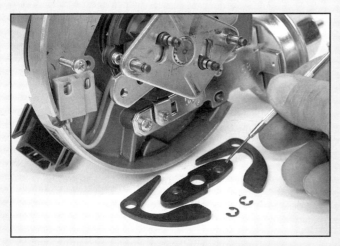

a. Remove the rotor, mechanical advance weights, and springs. On our Pertronix HEI there are two tiny C-clips that hold the center pivot in place. Remove the C-clips (don't lose them!) and remove the pivot.

b. Using a small, round punch, drive out the roll pin in the distributor gear and remove the gear and any shims from the distributor shaft.

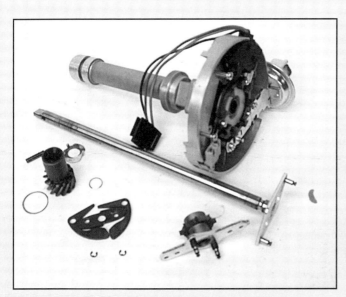

c. Pull the shaft out of the distributor and slide the weight pivot plate off the shaft. Note the location of the two pivot slots. The length of these slots determines the total mechanical advance that the distributor can achieve.

d. The Pertronix distributor mechanical-advance slot measured 0.560-inch long, which we shortened by approximately 0.160 inch to create a 0.400-inch long slot. This shortened the mechanical advance by 10 degrees. We ground the weld down, buffed it smooth, and reassembled the distributor with some white grease on the shaft.

13 Hot, Hot, Hot Cam

422 HP From A Budget Vortec Head and Hot Cam 355 cid Mouse
By Jeff Smith

Face it: Everybody's looking for The Deal—you know, the 400hp motor for $100. We're not there yet, but we're always searching for better ways to build a small-block and we think we've stumbled onto one that is plenty hot.

The key to a strong engine is the combination of great heads and an excellent camshaft. Improved airflow is usually accompanied by a hefty price tag, but check out this combo: Start with a late-model roller cam 350ci block. Bolt on a set of Vortec iron heads and a GM Performance Parts HOT hydraulic roller cam. Sprinkle on an aluminum GM Vortec dual-plane intake and you have an outstanding combination of all-new parts, with the heads, cam, and intake costing less than $800!

Our pal Tim Moore is building this exact package. But before you dive right into this deal, there are a few important details that you need to know that could save you money. The hinge pin to this plan is an '88-or-later 350ci roller-cam cylinder block. Chevy put these engines in Camaros, Firebirds, police cars, some Caprices, and 3/4-ton truck chassis also used in motor homes. These roller cam blocks differ from earlier blocks in several significant ways.

BLOCK TECH

Starting in '87, Chevy converted all small-blocks to a one-piece rear-main seal combined with an excellent one-piece oil pan seal. Chevy followed that in '88 with hydraulic roller cams in most V-8 engines. To accommodate the taller hydraulic roller lifter, Chevy increased the height of the lifter boss. This also required two small cast-in bosses in the middle of the lifter valley to mount the sheetmetal retainer, called the "spider," which retains each pair of lifters.

All this block information is given for several reasons. The GM Performance Parts HOT hydraulic roller is a great, affordable cam ($175 from Scoggin-Dickey Performance) and begs to be matched with a set of Vortec heads. This OEM-style roller cam is intended to work in '88-and-later roller cam small-blocks, because it requires a production thrust plate. However, according to our erudite engine enthusiast Kevin McClelland, you can use this cam in an early two-piece rear-main seal block. Trim the ears off the production roller-cam thrust plate and use it as a spacer between the cam gear and the block. Then employ a thrust button to control cam endplay. This will require using aftermarket hydraulic roller tappets, which are more expensive than the factory hydraulic roller tappets. This way, you could use your existing, older two-piece rear-main seal block.

Hot, Hot, Hot Cam

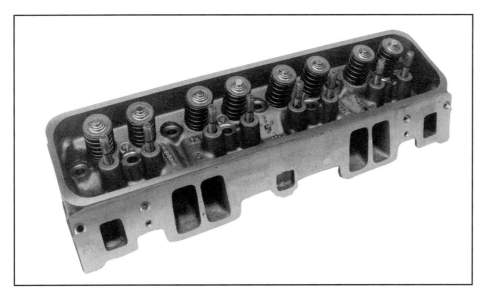

The heart of this budget Mouse is the combination of the GM Performance Parts HOT hydraulic roller cam and a set of Vortec heads. Combine these pieces with 9.6:1 compression, a 750-cfm carb, and a decent set of headers and you have 400hp potential.

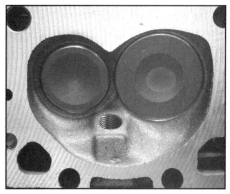

The heads use 1.94/1.50-inch valves, a 64cc combustion chamber, and 170cc intake ports.

Since Tim was building a whole new motor, he latched onto a used roller cam 350ci short-block that came out of an '89 Caprice police car. He disassembled it, had the crank machined and the rods rebuilt, and added a set of Federal-Mogul hypereutectic pistons. He also measured the deck height and had the block milled to establish a 0.005-inch piston-to-deck clearance. He also used the inexpensive GM Performance Parts cam drive assembly for the roller cam that is a screamin' deal at $40. Tim reused the stock hydraulic lifters, oil pan, and timing chain cover to keep the price down. To top it off, he bought a new oil pump from PAW and bolted the short-block together.

VORTEC HEAD MODS

Before Tim could bolt the Vortec heads on the engine, the castings required a slight tweaking to accommodate the larger HOT cam. The stock Vortec valvesprings cannot handle the HOT cam's 0.525-inch lift using the 1.6 roller rockers, so the tops of the valve guides must be machined shorter to clear the retainer. The spring pockets must also be opened up for the slightly larger-diameter springs. Tim also had the heads machined for ARP screw-in studs, but he did not use guideplates because he was using rail-style rocker arms.

The machine work for the springs configures the heads for the LT4 valvesprings that can handle the 0.525-inch lift and are also durability tested by Chevrolet to withstand performance abuse for hours on end. In fact, these heads and springs have individually been tested with Chevy's 300-hour wide-open throttle durability test and passed without failure. That's how good these components are.

Because the Vortec heads use a 64cc chamber, Tim used a Federal-Mogul hypereutectic 10cc dished piston to keep the compression ratio in line. Even with the 0.005-inch deck height and the Fel-Pro composition head gasket, the compression is still streetable at 9.6:1. That's at the high end of a streetable compression ratio with iron cylinder heads, but

Small-Block Chevy Engine Buildups

CAM SPECS

The advantage of a roller camshaft is its ability to generate a more aggressive lobe profile. This allows the cam to push the valves open at a faster rate, which means you can combine an aggressive valve lift with short-duration figures. This pumps up the power across the entire powerband of the engine. We've listed the HOT cam, as well as the RamJet hydraulic roller cam. We've also listed a GM hydraulic flat-tappet cam with similar duration figures at 0.050-inch tappet lift to illustrate the lift advantage enjoyed by the hydraulic roller cam. You can see that the HOT cam enjoys a significant lift advantage over both the RamJet and the flat-tappet cam.

	Duration advertised (degrees)	Duration @ 0.050 (degrees)	Lift 1.5 (inches)	Lift 1.6 (inches)	Lobe sep. angle (degrees)
HOT					
Int.	279	218	0.492	0.525	112
Exh.	287	228	0.492	0.525	
RamJet					
Int.	288	196	0.431	0.459	109
Exh.	308	206	0.451	0.480	
Marine					
Int.	-	214	0.442	0.470	112
Exh.	-	224	0.465	0.496	

Using the HOT cam also requires modifying the valve seats for larger valvesprings. You must also trim the inside diameter of the valve guides to clear the inside diameter of the springs.

combined with this camshaft should make excellent torque and not suffer from detonation.

TESTING

We trucked this iron-headed small-block out to Westech Performance and let John Baechtel and Steve Brulé abuse this Mouse on their SuperFlow dyno. Moore elected to use an old Stinger electronic ignition distributor to trigger the MSD 6 box used on Westech's dyno. This also required a special melonized GM distributor gear to ensure the gear would live when used with the GM Performance Parts HOT hydraulic roller camshaft. This gear is not cheap, but it does work. As for exhaust, we bolted on a set of Hooker 1 5/8-inch headers along with a 2 1/2-inch exhaust system plumbed to a pair of Flowmaster mufflers. For the final piece of the puzzle, we added a Barry Grant Speed Demon 750 me-chanical secondary carburetor to ensure that the carburetor would not restrict power.

Since this was a newly rebuilt engine, we pre-oiled the assembly and set the static timing to allow it to fire up immediately. Moore set the total timing at 34 degrees and then allowed the engine a short 20-minute break-in session. Finally, with a nice leak-free engine, we loaded up the fuel cell with 92-octane 76 gasoline and pulled the handle.

For this initial test, we used the GM Performance Parts dual-plane Vortec intake manifold. The first thing we wanted to see was how well the engine idled. The motor responded with an amazing 14 inches of manifold vacuum at 850 rpm, which means this could easily support vacuum-operated accessories like power brakes. The first few pulls looked slightly rich, so Brulé pulled two jet sizes out of the carb, but first we had to do a minor tune-up on the driver-side header.

One reason the Vortec heads work so well is because of the superior chamber design and the fact that the engineers moved the spark plug closer to the exhaust side of the chamber. Vortec heads are a straight-plug design that sometimes can cause header clearance problems. In the case of the Hooker headers, the No. 5 spark-plug boot was pushed too close to the header tube and melted. This required us to heat and dent the

Hot, Hot, Hot Cam

GM Performance Parts sells both a dual plane (top) and a single-plane intake manifold for the Vortec cylinder heads. The dual plane is the better choice for a street engine since that will pump up the torque curve, making the engine more streetable.

Scoggin-Dickey supplied an entire valvespring kit, which is also in the GM Performance Parts catalog. The spring kit includes LT4-style springs, lightweight retainers, and the proper keepers to install the springs on the Vortec heads.

The production GM roller-cam block uses taller lifter bores to support the hydraulic roller tappets. Roller tappets must not rotate in their bores, so small cast retainers tie pairs of the rollers together. This large stamped-steel "spider" keeps the retainers in place.

header to create enough air gap to protect the spark plug–wire boot.

Once we had all eight cylinders firing with slightly leaner jets in the Demon carburetor, the 355 pulled up some impressive numbers. Even at 2,500 rpm, the 350 managed 359 lb-ft of torque with a max torque of 423 lb-ft at 4,400. Peak horsepower came in at 412 at 5,600 rpm, creating a somewhat narrow powerband of only 1,200 rpm, but looking more closely at the numbers reveals the engine is punching out 400 lb-ft from 3,500 all the way through 5,300 rpm.

While these numbers are respectable, we also wanted to try the Edelbrock Super Victor single-plane intake for Vortec heads. We expected the single plane would trade power below peak torque for more power above peak torque. Generally, this is not a good trade for street engines, but Moore also plans to use nitrous on this engine and felt that the single-

Small-Block Chevy Engine Buildups

Factory '88-and-later roller-cam engines use a different mounting face. Use the GM cam drive kit, which sells for $40.

The HOT cam also uses a normalized iron distributor gear that does not require a bronze gear.

plane intake would work better with a plate nitrous system.

Changing intakes was a snap, and once again we were working this small-block. After a few minor tweaks, the numbers looked very good, but we also decided to try adding a 1-inch spacer as well as an additional 1/2-inch spacer to simulate the nitrous plate. We also experimented with reducing the timing to 32 degrees total instead of the previous test's 34 degrees.

Not surprisingly, the 355 did sacrifice some power down low, but actually very little. What was more surprising was the 422 hp (a 10 hp gain) at 5,700 rpm we saw over the dual plane. The trade-off was torque, where the single-plane intake managed still to muster 418 lb-ft of torque at 5,000. The peak horsepower rpm points were almost identical, yet the single plane managed 10 more horsepower and created a virtually identical average torque figure.

CONCLUSION

What this all comes down to is a great small-block. The dual plane made at least 360 lb-ft of torque from 2,500 to 6,000 rpm, and the single plane was only slightly behind. But the single-plane version cranks up the horsepower with well over 420 ponies.

We didn't try this, but with only 32 degrees of total timing, it's entirely possible that despite its 9.75:1 compression ratio, this engine might run just as hard on 87-octane gas as it did on 92-octane gas. This is a direct result of a tight quench area, an excellent combustion chamber design, and careful attention to assembly. Add the 150–175 hp nitrous system that Moore wants to run and we're talking about a thumper 11-second street car. Even with a Demon carb, headers, an ignition, and a nitrous kit, you still could build this complete engine setup for under $4,000, and that's if you bought everything new. That's a winning combination any way you look at it.

Hot, Hot, Hot Cam

All '87-and-later Chevy blocks come with a one-piece rear-main seal. This requires the use of a late-model one-piece rear-main seal crank. Starting with the '88s, most passenger-car engines converted to hydraulic roller cams that required a spider. This is a truck block where the spider mounting bosses are not drilled and tapped, because the truck engine used a flat-tappet cam. These can be easily drilled and tapped to mount the spider for a hydraulic roller cam.

Tim also used a set of rail-style 1.6:1 roller rockers. The small rails locate the rocker arm over the valve.

The GM Performance Parts dual-plane intake manifold combined with the Demon carburetor worked extremely well on the dyno, making 422 lb-ft of torque and 412 hp at 5,600. These are excellent numbers for a pump-gas street engine.

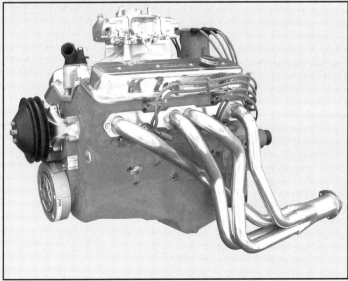

As tested, we configured the Vortec-headed small-block with a set of 1 5/8-inch headers, the single-plane intake, a set off Moroso polished valve covers, plug wires, and a trick distributor cap. All tests were performed on Westech's SuperFlow dyno corrected to a standard 29.95 barometer, 60 degrees F, and no humidity. Note that the dual plane made slightly more torque below peak torque with less horsepower.

Small-Block Chevy Engine Buildups

The dual-plane manifold (bottom) offered increased torque, while the Edelbrock single plane (top) made more horsepower. Yet even as the power curves varied, the average torque remained within 1 lb-ft over the entire power curve.

The Demon 750-cfm mechanical-secondary carburetor performed flawlessly in this test. It only required minor jetting changes to optimize the power in both intake manifolds.

This test required a full street exhaust system, so we bolted on a set of Hooker 1 5⁄8-inch headers along with a pair of 2 1⁄2-inch exhaust pipes clamped to a pair of Flowmaster mufflers.

Swapping intake manifolds was easy since the Vortec intake is only held in place with eight bolts. The hardest part is replacing the distributor, and that's not difficult at all.

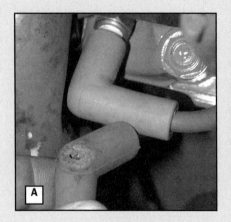
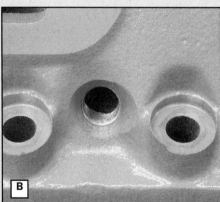
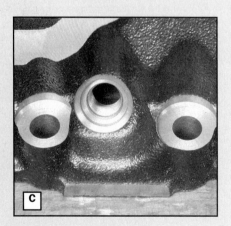

Note the spark plug–wire boot we melted in the first test (a). Most Chevy heads place the spark plug squarely between the head bolts (b). The Vortec head moves the spark plug closer to the exhaust valve (c), which can cause clearance problems for some headers.

Hot, Hot, Hot Cam

Dyno Test

RPM	HOT 1 Dual-plane Intake TQ	HP	HOT 2 Single-plane Intake TQ	HP	Difference TQ
2,600	358	177	363	179	+5
2,800	364	194	364	194	0
3,000	371	212	370	211	–1
3,200	381	232	376	229	–5
3,400	399	258	389	252	–10
3,600	412	282	402	276	–10
3,800	418	303	408	295	–10
4,000	419	319	413	315	–6
4,200	419	335	415	332	–4
4,400	423	354	415	347	–8
4,600	422	369	416	364	–6
4,800	419	383	417	381	–2
5,000	415	395	418	397	+3
5,200	407	403	412	408	+5
5,400	399	410	407	418	+8
5,600	386	412	394	420	+8
5,700	379	411	389	422	+10
5,800	372	410	380	420	+8
6,000	357	408	365	417	+8
Avg. Torque:	396.8 lb-ft		396 lb-ft		

PARTS LIST

All pieces listed below are from GM Performance Parts.

COMPONENT	PN
Vortec cylinder heads, pr.	12558060
HOT cam	24502586
HOT cam kit*	12480002
H.O. lifter kit**	12371042
Roller rocker arms, 1.6	12370839
Valvespring kit	12495494
Retainer kit	12495492
Keeper kit	12495503
Pushrod kit, roller, 2nd design	12371041
Timing chain kit	12371043
Intake, Vortec, dual plane	12366573
Intake, Vortec, single plane	12496822
Bolt kit, intake, stainless	SD12550027S
Intake gasket, Vortec	12529094
Wire harness, spark plug	12496806
RamJet hyd. roller cam	14097395
Oil pan, Camaro 1 pc rear-main	12557558
Gasket, valve cover (each)	10046089
Stock hydraulic roller lifters, PAW	PAW-2148HR

*The HOT cam kit contains the cam, 1.6:1 roller rockers, LT4 valvesprings, lightweight retainers, keepers, and shims.

**The H.O. lifter kit contains hydraulic roller lifters, guides, special-length roller cam pushrods, the sheetmetal "spider," bolts, and washers.

14 Son of Muscle Mouse

Our 515 HP 406 Dart Is Reborn
By Jeff Smith
Photography by Ed Taylor

In a time not so long ago, CHP built a motor we called the Muscle Mouse. With a stock iron 400 block, Dart 215cc Iron Eagle heads, a Scat cast crank, and 4340 steel rods we had a winner on our hands. On Ken Duttweiler's dyno, the small-block strongarmed its way to 455 hp at 5,500 and 498 lb-ft of torque at a streetable 3,800 rpm. We stuffed the motor in Bob Moore's Camaro and took it to the track. That's when it proceeded to eat itself alive.

The autopsy revealed that the engine had been improperly balanced, because we did not include the roll pin in the rear crank flange, and the externally balanced flexplate was positioned one bolt hole off during the balance procedure. When the flexplate was installed correctly in the car, the slight imbalance quickly destroyed the mains, killing the crank, block, and rods.

After the memorial service, we vowed to build another small-block. This chapter is all about the Son of Muscle Mouse. In the gestation period, Ventura Motorsports engine builder Ed Taylor and the CHP staff decided to build this version with an emphasis on more muscle.

THE TECHNOLOGY

Foundations are the key to horsepower. We plan on supercharging this rascal at a later date, so we called Dart Machinery for one of the Little M blocks. With thicker siamesed cylinder walls and a 0.625-inch minimum deck thickness, the Dart Little M is one stout block. The pan rails are spread 0.400 inch while retaining the more popular two-piece rear-main seal. If desired, this block could be built as large as 455 ci using the 0.300-inch taller 9.325-inch deck height.

While we were sorely tempted to crank the displacement with a longer stroke, we stayed with the stock stroke but pumped the strength with a Scat 4130 steel forged crank. Matching the crank is a set of Scat 4340 steel H-beam–style rods pinned to Sportsman Racing Pistons (SRP) forged pistons with 21cc-dished pistons to limit the compression to 9.8:1. Along with the Speed-Pro 1/16-inch rings and the Federal-Mogul bearings, we also included a complete Moroso lubrication system.

Our original 406 used Dart Iron Eagle 215cc heads, but we decided to step up to a set of Dart PRO-1 aluminum CNC-ported heads. These heads retain the standard 23-degree valve angle sitting in a 66cc CNC-machined combustion chamber using a 227cc intake port. The heads are finished off with a set of 2.08/1.60-inch valves and 1.550-inch diameter roller cam springs and titanium retainers.

CHP's flow testing revealed nearly 300 cfm of flow at 0.600-inch lift on the intake side, but more importantly, these heads crank out some impressive flow numbers at 0.300- and 0.400-inch lift. It is these mid-lift flow numbers that will support a strong torque curve as well as excellent peak power.

In keeping with our Dart theme, we also added Dart's single-plane intake manifold mounted with a Barry Grant Race Demon 825-cfm double-pumper carburetor. The Race Demon also features removable venturis so we can experiment with various sizes to more accurately match the carb size to the engine.

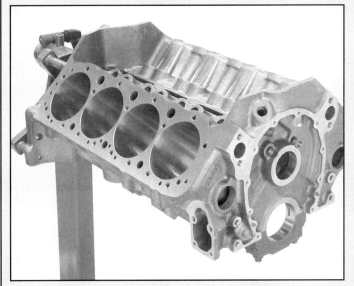

We have big power plans for this engine, so that called for a strong block. Dart's Little M iron block offers thick cylinder walls, a sturdy deck, and massive steel main caps.

Nothing less than Sportsman Racing forged dished pistons would do for this motor. The dish keeps the compression down to 9.8:1. The rods are Scat H-beam 4340 forgings with ARP bolts.

Scat also supplied the 4130 forged-steel internally balanced 3.75-inch stroke crank held in place with ARP fasteners.

Small-Block Chevy Engine Buildups

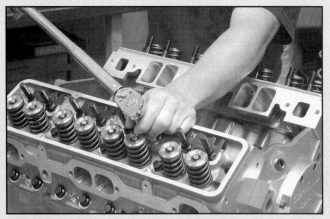
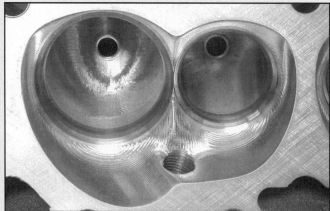

The Dart PRO-1 CNC-ported heads deliver the air and fuel past 2.08/1.60-inch valves and a fully CNC-machined 66cc combustion chamber. We added matched Comp Cam springs to these heads to ensure no valve float problems.

To ensure adequate lubrication, we also added a Moroso oil pump with its pickup factory-welded in place along with a matching performance oil pan with trap doors to maintain the oil in the sump.

We decided to step up to a Comp Cams Xtreme Energy mechanical roller cam for this descendant of Muscle Mouse.

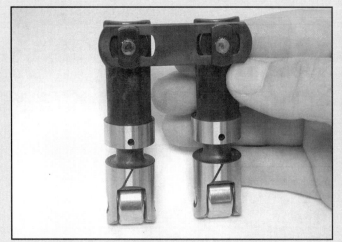

We also opted for the pressure-slotted Endure-X roller lifters that ensure pressure lubrication to the cam-lifter interface.

With a roller cam, it just makes sense to complete the system with a high-quality set of Pro Magnum roller rockers. We started with a set of 1.6 rockers to pump the valve lift to 0.614 inch on the intake and 0.620 inch on the exhaust.

Son of Muscle Mouse

Dart also makes a killer single-plane air gap–style intake that we couldn't resist. To this we mounted a Barry Grant Race Demon 825-cfm double-pumper carburetor.

ATI makes what many engine builders feel is the best damper on the market. We combined it with a Tavia pointer to ensure accurate ignition timing.

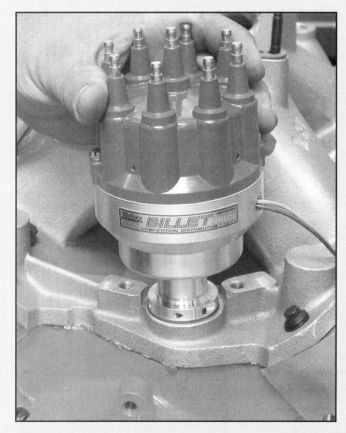

When you're shooting for over 500 hp, ignition plays an important role in lighting the fire. We opted for a complete Mallory ignition system including a Hy-Fire VI box, electronic distributor, and coil.

Flow Chart

The following is a flowchart of the out-of-the-box Dart CNC-ported PRO 1 aluminum heads. The heads were tested at Westech Performance Group by Steve Brule and will be included in our latest update of flow-tested cylinder heads. Note the strong exhaust flow numbers.

Dart PRO 1 CNC-ported
2.08/1.60-inch valves
66cc chamber
227cc intake port

Valve Lift	Intake	Exhaust Open	Exhaust w/pipe
0.050	36	26	23
0.100	71	68	62
0.200	144	112	103
0.300	193	147	139
0.400	236	189	185
0.500	270	215	211
0.600	292	224	220
0.700	293	229	224
I/E @ 0.400	80%		

Small-Block Chevy Engine Buildups

DYNO TESTING

Here's where it gets serious. All dyno-testing was performed at Duttweiler Performance. Typically, larger headers make more peak horsepower, but note how the curve changed with the larger headers, increasing torque between 3,200 to 3,600 rpm. Because the exhaust port is so efficient, the smaller headers over-scavenged in this area. Test 1 used 1 5/8-inch Hedman headers on the 406. Test 2 changed to a set of 1 7/8-inch Hedman headers to test the engine's peak horsepower potential. Idle vacuum is a bit radical at 5 inches at 950 rpm.

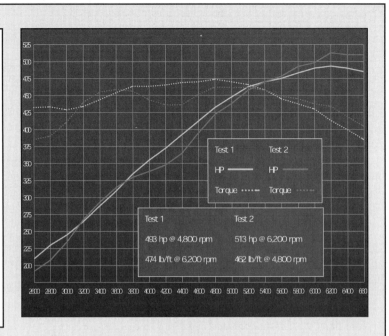

Test 1	Test 2
493 hp @ 4,800 rpm	513 hp @ 6,200 rpm
474 lb/ft @ 6,200 rpm	462 lb/ft @ 4,800 rpm

PARTS LIST

Component	Mfr.	PN
Cylinder block, iron	Dart	PN 31132211
Heads, PRO 1 CNC	Dart	PN 11971143
Intake manifold	Dart	PN 42311000
Valve covers	Dart	PN 68000010
Crankshaft, 4130 forged	Scat	PN 400375541
Connecting rods, 4340	Scat	PN 3506000R
Pistons, 21cc dished	SRP	PN 139625
Rings, 1/16-inch	Speed-Pro	PN R9346-35
Bearings, main	Federal-Mogul	PN 140M
Bearings, rod	Federal-Mogul	PN 8-7100CH
Oil pump, w/pickup	Moroso	PN 22135
Oil pan	Moroso	PN 20191
Water pump	Moroso	PN 63500
Gasket set	Fel-Pro	PN 2802
Pan gasket, one-piece	Fel-Pro	PN 1881
Camshaft, roller	Comp Cams	PN 12-772-8
Camshaft, entire kit	Comp Cams	PN CL12-772-8
Lifters, roller	Comp Cams	PN 888-16
Timing set	Comp Cams	PN 7110
Rocker arms, 1.6	Comp Cams	PN 1105-12
Pushrods	Comp Cams	PN 7972-16
Valvesprings	Comp Cams	PN 977-16
Retainers, steel	Comp Cams	PN 740-16
Valve locks	Comp Cams	PN 611-16
Thrust button	Comp Cams	PN 200
Timing cover, 2-piece	Comp Cams	PN 210
Balancer	ATI	PN 917785
Timing pointer	Tavia	PN 02344
Hy-Fire VI ignition box	Mallory	PN 685
Distributor	Mallory	PN 8448215
Coil	Mallory	PN 29440
Plug wires, 8mm	Mallory	PN 937
Carb, Demon 825 cfm	Barry Grant	3423010
Fuel pump	Carter	PN M4891
Headers, 1 5/8-inch	Hedman	see catalog
Pulleys and brackets	Zoops	see catalog

CAM SPECS

The following are cam specs for the Comp Cams mechanical roller cam that we used in the Son of Muscle Mouse. Based on the dyno results, we feel that a single-pattern version of this cam might make more power, especially in the low- to midrange torque band. A single-pattern cam would have the same duration and lift numbers on both the intake and exhaust lobes.

Camshaft	Duration advertised (degrees)	Duration @ 0.050 (degrees)	Lift (inch) w/1.5 rockers	Lobe Sep.
XR-286R				
Xtreme	I 286	248	0.576/0.016" lash	110
Roller	E 292	254	0.582/0.018" lash	

Gladiator vs. Muscle Mouse

A Comparison of Two 500 hp Street Engines
By Jeff Smith
Photography by Jeff Smith

I love your magazine, especially the engine buildups. After buying the Jan. '02 issue, a question popped into my head. You built a 406 small-block ("Son of Muscle Mouse") with heads that flow 300 cfm and an intake manifold that costs around $350 to make 515 hp. Now I remember a 355ci small-block that you featured called the Gladiator (Feb. 2000) that was 51 ci smaller than the 406, used AFR 195 heads, flowed in the 260-cfm range with an intake that costs about $160, and made 524 hp. That sounds kinda weird to me. Was that 406 running on four cylinders? Both engines have roughly the same compression, camshaft, and carburetor. What do you guys think? John Kobik, via e-mail

As you can see from this letter, our readers constantly challenge us, which is what makes this job so interesting. John Kobik's letter asks a great question that we felt needed to be answered completely.

Power is what everybody wants. If all you did was look at the peak horsepower numbers, it would appear that the Gladiator engine beats the much more expensive 406. This is especially true when you consider that the Gladiator is 51 cubic inches smaller. But this overlooks the torque created by the 406. So we decided to take an in-depth look at both of these engines and the power they produce.

The first thing we did was compare the power curves. Right away, it's clear the 406 makes almost 50 lb-ft more torque at 3,500 rpm, which is the lowest common rpm point. Of course you would expect the larger motor to make more torque, but this is a ton of power.

As the rpm climbs, the power differential begins to favor the 355, and at its peak, the Gladiator makes 14 more horsepower.

Looking at the average numbers, the 406 makes more average torque, but the Gladiator makes more average horsepower. While 14 more horsepower sounds impressive, it's average power that eventually moves the car down the track. So we plugged both power curves in the Racing Systems Analysis Quarter Pro computer simulation program. We simulated a 3,500-pound Chevelle with a TH350 automatic, a 3,000-stall converter, 3.73 gears, and a shift point of 7,000 rpm.

Here's where it gets interesting. The simulated quarter-mile passes for both engines in the same theoretical car were within 0.05 second of each other. The 355 Gladiator ran an 11.36/119.30-mph pass with a 60-foot time of 1.69 seconds. The 406 Muscle Mouse motor ran a slightly better 1.65-second 60-foot time and pulled off an 11.31/118.90-mph run. In both cases, we optimized traction with a big enough tire to eliminate tire spin. The 406 got to the finish line first, but

Small-Block Chevy Engine Buildups

"In the red corner, weighing in at 355 inches with a reach of 524 hp—the Gladiator!" With AFR 195 heads, a mechanical-roller cam, and a big single plane, this Westech-built Mouse is one of the strongest street-style small-blocks we've tested.

THE ENGINES

We also looked more closely at the parts used in both these engines. Amazingly, these engines are far more similar than they are different. The Gladiator started life as a ZZ4 engine that Westech's owner John Baechtel converted to forged Federal-Mogul 10:1 compression pistons and rings for increased durability. He also added a set of out-of-the-box Air Flow Research 195cc aluminum heads. Then, over the course of years of beating on this small-block, the engine has witnessed dozens of camshaft, intake manifold, carb, and header swaps in search of more power.

In the test we included here, the Gladiator went into battle with a big-port Victor single-plane intake, an 850-cfm Speed Demon carb, and a set of 1 3⁄4-inch street headers pumping through a set of Flowmaster mufflers. We've also included the complete

the 355 motor had the stronger top-end charge. In reality, these two engines would run very similarly at the track despite the 406's apparent lack of horsepower.

This illustrates the significance of torque in the overall power curve and its importance in acceleration. In this simulation comparison, we left the converter exactly the same. However, had we subjected the 355 to a lower stall converter where that motor makes less torque, the 406 would have pulled the smaller motor with both a quicker e.t. and higher trap speed. This points out how important it is to tune your chassis and drivetrain combination to take advantage of the engine's power curve. In this case, the Gladiator engine would really benefit from a four- or five-speed manual trans leaving at 3,500 rpm or higher.

"In the blue corner, weighing in at 406 inches with a punch measuring 513 hp—the Muscle Mouse!" These larger motors are notorious for torque, which is this motor's strength

Gladiator vs. Muscle Mouse

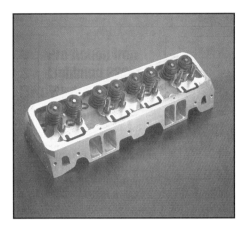
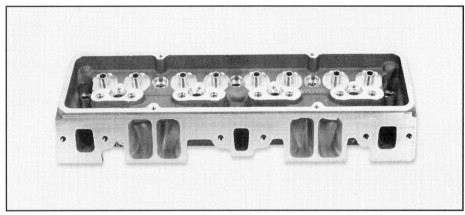

Cylinder heads are the single most important component when it comes to making horsepower. John Baechtel chose a set of Air Flow Research 195cc aluminum heads (left) for the Gladiator engine. For the Son of Muscle Mouse 406, we chose a set of Dart Pro 1 CNC 227cc heads (right), which offer outstanding flow potential.

Both engines also sported Comp Cams mechanical-roller cams with 1.5:1 roller rockers. Mechanical rollers offer big valve-lift numbers while keeping the seat duration relatively short compared to a flat-tappet cam.

The 406 sported a Dart single-plane intake (right) and a Race Demon 825-cfm carb, while the 355 featured a similarly big Edelbrock Victor intake using a Street Demon 850-cfm carb and a 1-inch tall carb spacer.

cam specs on the Comp Cams solid-roller cam (see "Cam Specs"). As you can see from the chart, the Gladiator 355 enjoyed a slightly longer-duration cam. Combined with its smaller displacement, this is a big reason why the Gladiator made more power than the 406.

On the 406ci side, the Muscle Mouse motor enjoyed a much larger intake port with the 227cc Dart Pro 1 heads, which had also been treated to a full CNC porting. Compression with the 406 was slightly lower at 9.8:1, and the Comp Cams mechanical roller was also slightly shorter in duration and lift compared to the Gladiator grind. We used a similar large-port Dart single-plane intake, and the carb on the 406 was also a similarly sized Demon 825-cfm Race Demon fuel mixer. One difference was that we used a set of 1 7/8-inch headers on the 406 flowing through a pair of Borla stainless mufflers.

CHANGES

The power curve we published for the Gladiator was the culmination of dozens of dyno pulls on that engine, and it was one of the best power curves this engine ever generated. On the other hand, the numbers on the 406 represented the first test of this engine. We're not using that as an excuse, but we also now have a plan to improve the power curve.

For example, after studying the flowbench data, it's clear that the Dart Pro 1 CNC heads demand a single-pattern camshaft where the exhaust duration and lift are the same as the intake. Combined with the outstanding 0.100-inch exhaust-flow numbers, a shorter-duration cam will help prevent the over-scavenging of the cylinder during the exhaust event. We think this hurt the 406's final power output.

It's clear that the 1 7/8-inch headers were too big for this combination. We also tested a set of 1 5/8-inch headers and at 2,800 rpm the smaller headers were worth an additional 43 lb-ft of torque. The downside of the smaller headers is a 20hp loss at horsepower peak.

If you look at these changes all together, we're working mainly on the exhaust side of the engine. The smaller 1 3/4-inch headers will improve torque in the midrange, while the shorter-duration cam should help the peak by reducing the scavenging effect of the longer duration. We're not sure what all this will be worth, but we're hoping for 10 to 15 horsepower along with an equal amount of torque.

All this illustrates the importance of matching components. We can conjecture and guess, but ultimately it comes down to trying several different combinations in search of more power, but even then it won't be even close to the ultimate power this motor can make. That's what makes all this engine building and testing so much fun. Just when you begin to think you might have hit the power limit, somebody comes along with an idea and raises the power bar another notch.

Flow Figures

The following are flow numbers for both of the cylinder heads used in this comparison. We have reproduced the entire flow test, but keep in mind that both cams limited intake valve lift to roughly 0.580 inch. This means that port flow performance above 0.600-inch valve lift is irrelevant. Another important point is that the AFR head uses a 30cc-smaller 195cc intake port than the 227cc Dart Pro 1 does. Note the substantial difference in exhaust-to-intake (E/I) flow percentages. The Dart offers a much stronger exhaust port that tells us we should use a single-pattern cam in the 406.

Valve lift	AFR 195 CNC Intake	Exhaust w/pipe	E/I	Dart Pro 1 CNC Intake	Exhaust w/pipe	E/I
0.100	71	31	43%	71	62	87%
0.200	144	67	46%	144	103	71%
0.300	208	121	58%	193	139	72%
0.400	244	157	64%	236	185	78%
0.500	262	188	72%	270	211	78%
0.600	261	202	77%	292	220	75%
0.700	—	211	—	293	224	76%
Avg. E/I			60%			77%

Cam Specs

	Duration @ 0.050 (degrees)	Lift (inches)	Lobe Separation angle (degrees)
Gladiator 355ci	I –254 E –260	0.583 0.585	110
Muscle Mouse 406ci	I –248 E –254	0.576 0.582	110
Proposed Cam for 406ci	I –248 E –248	0.576 0.576	110f

Both of these cams are Comp mechanical roller cams. As you can see, the Gladiator cam offers 6 degrees more intake and exhaust duration along with slightly more lift.

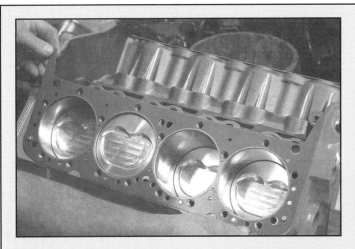

Another critical criterion when shooting for 500 hp and beyond is proper quench. The Sportsman Racing Pistons (SRP) 406 forging uses a D-shaped dish to keep compression below 10:1, but it also uses a large flat portion to add turbulence in the chamber.

The 406 enjoys a larger 4.155-inch bore, which helps airflow, but it also suffers from a longer 3.75-inch stroke which, along with larger bearings, means more internal friction than the shorter-stroke 355. This is one reason why 400ci engines do not make as much horsepower per cubic inch as shorter-stroke engines.

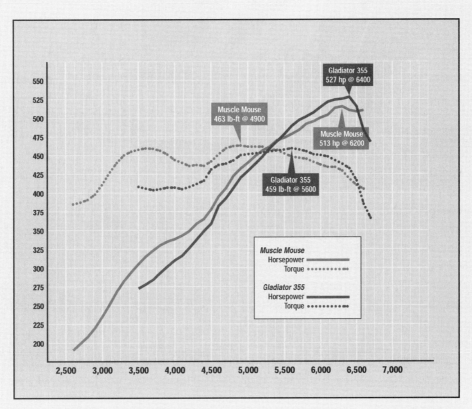

The two engines were tested on different dynos, which could also account for a slight difference. We intend to put the 406 back on the dyno with a shorter exhaust lobe cam to see if we can improve its power output.

16 Battle of The Small-Block Strokers

377 vs. 383 Slugfest
By Jeff Smith
Photography by Ed Taylor

There are many famous duos in American history. There's Laurel and Hardy, Batman and Robin, Abbott and Costello, Fred Astair and Ginger Rogers, the Green Hornet and Kato, and—who can forget—frank and beans. When it comes to small-block Chevys, we decided to create our own famous duo with a pair of small-blocks intended to test the timeless bore-stroke question. For street enthusiasts, bigger is always better and most feel that displacement will always win the king-of-the-hill contest. But find an informed engine builder and he will tell you that this isn't always the case. Sometimes a smaller displacement engine with a bigger bore can make more horsepower.

There are thousands who subscribe to the classic small-block 383 power plan as the easiest and least-expensive path to increased stroker power. Then there are the proponents of the 377 as the very capable, bore and stroke combination. We decided the best way to determine who's best was to build two otherwise identical engines and put 'em to the test. The winner would be the engine that makes the best overall power.

In honor of these famous duos, engine builder Ed Taylor has dubbed our pair of battling small-blocks Donnie and Marie. The 377 is all rock 'n' roll (that's Donnie), while the 383 is a little more country (that's Marie). This story will detail the construction of these two engines. In most ways, they are very similar, but in certain important ways, they are significantly different. This story will detail the differences in case you'd like to duplicate one of these healthy small-blocks for yourself.

THE COMBATANTS

Before we get into the details, let's take a close-up look at our internal-combustion contenders. In the near corner, we have the 383 representing the long-stroke/small-bore plan. A 383 small-block consists of a 4.00-inch-bore 350 block bored 0.030-inch oversize and employing a longer 400-style 3.75-inch-stroke crankshaft compared to a 350's stock 3.48-inch stroke. This is an easy bore/stroke combination to assemble since all you have to do is machine the 400 crank's large 2.65-inch-diameter main-bearing journals down to the 350's smaller 2.45-inch diameter. An even easier alternative is to buy an aftermarket 383 crank that is already machined to the 350 main-journal dimension.

In the far corner is our 377ci bruiser. This short-stroke/big-bore version drops the shorter 350ci-style 3.48-inch-stroke crank in a 400ci block. Bore the cylinders 0.030-inch oversize to 4.155 inches and you have a 377ci engine. The only snag here is that the 400 block requires either thicker, custom main bearings, main-bearing spacers, or a custom-ground

Battle of the Small-Block Strokers

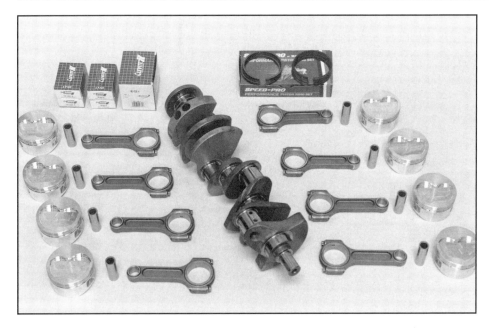

The heart of each engine is the rotating assembly. For both engines, we went with Summit's Pro Line 4340 steel crank and rod assemblies. The cranks are forged and accurately machined so that the only additional work needed was to balance each assembly. The 377 used JE forged pistons, while the 383 employed off-the-shelf SRP pistons. Speed-Pro rings and Clevite bearings round out the bottom end.

The 350 block had to be relieved near the bottom of the bores to clear the larger-stroke crank. Be very careful here since it is easy to hit a water jacket, as we did. This required meticulous cast-iron welding to repair. Be careful, and avoid that additional hassle.

Since we were using 6-inch Summit Pro Line rods for both engines, this stuffed the wristpin up into the oil ring on the longer-stroke 383 engine. This photo shows the support ring used to stabilize the oil ring in the SRP 383 piston.

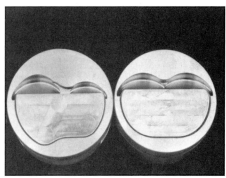

Two different pistons also required slightly different dishes to establish the exact same compression ratio. The larger-bore 377 piston (left) used a slightly larger dish.

crank to make up the difference between the 350 crank's 2.45-inch main journals and the 400 block's larger 2.65-inch main journal diameter. This main-journal difference can make this combination more expensive to build. Rod-journal diameters for both engines are the same at 2.10 inch.

The power theory behind the 377 is that the larger 4.155-inch bore unshrouds the valves, which should allow the heads to breathe better. This should result in more horsepower. The advantage of the 383 is that its longer stroke crank should produce more torque than a 377. Most 383 proponents will admit that a 377 can make slightly more horsepower, but the added expense doesn't warrant the marginal power increase. Besides, the 383 builders say, why give up all those cubic inches by destroking a 400? We decided to build two engines to find out.

THE BUILDUP

Ed Taylor of Ventura Motorsports took on the task of building these two engines and used as many off-the-shelf parts as possible. The first

Small-Block Chevy Engine Buildups

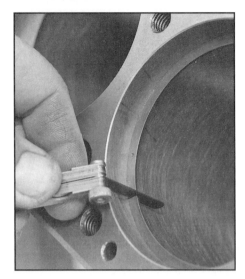

A good engine builder also carefully sets the ring endgaps to ensure consistency. We used 0.005-inch oversize Speed-Pro rings to carefully set the ring endgaps.

Taylor also spent considerable time ensuring all the bearing clearances were dead-on accurate. This requires hours of effort to measure everything, but it eliminates clearances as a difference between the engines. This is the 383 engine that uses cast-iron four-bolt main caps.

Here, Taylor torques the Summit Pro Line steel crank into the 377 motor. Note the center three Gellner Engineering four-bolt steel main caps.

Another important clearance that some backyard engine builders forget is the crankshaft endplay. With the rear thrust main cap torqued in place, Taylor measures the clearance between the crank thrust flange and the main-bearing thrust surface. This is the 383 engine.

variable we faced was also the one without a right answer. By keeping the rod length the same at 6.00 inches for both engines, this introduced a variable in the rod-length-to-stroke ratio (rod length divided by stroke). This creates a favorable rod-length ratio of 1.72:1 for the 377 and a slightly less favorable 1.6:1 ratio for the 383. The other way to go would have been to use 5.7-inch rods in the 377 and 6.0-inch rods in the 383, which would have equaled the rod-length-to-stroke ratios, but then the engines would not be exactly the same. We decided to use the longer 6.0-inch rods in both engines since the longer rods are more favorable.

After finding suitable 350 and 400 production blocks (which for the 400 wasn't easy), the next step was the rotating assemblies. After much research, we decided to use Summit's Pro Line series of cranks and connecting rods to swing the arms in both engines. The Pro Line crankshafts are high-quality 4340 steel cranks that have been tested up to 700 hp and 8,000 rpm—far beyond what we had in mind. The key to avoiding the spacer-bearing routine is to use a crank—like Summit's Pro Line piece—that is ground specifically with a 3.48-inch stroke but with the larger 400-style main bearings and

Battle of the Small-Block Strokers

Taylor used ARP's aluminum crank protectors to guide the piston and rod assembly over the crank. This guide protects the crank from nasty scratches.

This is the most critical torque procedure in the entire engine-assembly process. Taylor uses an ARP rod-bolt stretch gauge to accurately set rod-bolt stretch to ensure proper clamping load on the rod.

We also used ARP's tapered ring compressor to squeeze the rings enough to slip the pistons into the bores.

6.0-inch rods. This made the 377 literally a drop-in. The 383 was another shoe-in since Summit also offers a 3.75-inch-stroke crank with 350 mains for either 5.7- or 6.0-inch rods.

It would be dumb to bolt on a set of stock rods to these high-quality cranks, so we chose a set of excellent Pro Line 4340-forged steel 6.0-inch connecting rods to complement the cranks. These rods use excellent 8740 ARP capscrews instead of bolts and nuts and also employ guide pins that accurately locate the cap on the rod. These rods are far superior to stock rods and eliminated any concern about durability for our two test engines.

For pistons, we selected Sportsman Racing Pistons (SRP) forgings for the 383 and JE pistons for the 377. Both employ a dished design that retains the quench area of the piston to keep the engines inside the 9.5:1 pump-gas compression zone. The dished, 6.0-inch rod-length piston requirement for the 377 is not an off-the-shelf item, necessitating custom-ordered pistons from JE. This is another advantage the 383 enjoys since the 383's pistons are off-the-shelf Sportsman Racing Piston (SRP) items. Both pistons utilize a 1/16-, 1/16-, 3/16-inch Speed-Pro ring package that reduces friction while still maintaining excellent cylinder pressure and oil seal. Clevite 77 also supplied the rod and main bearings to support both Summit Pro Line steel crankshafts. To finish off the bottom end, Taylor added matching Moroso oil pans, windage trays, and oil-pump assemblies.

Since this test was an evaluation of the power difference in bore and stroke, we employed the same camshaft for both engines. In discussions with Comp Cams, we wanted to maximize the potential of these

Small-Block Chevy Engine Buildups

Next, Taylor used an ARP stud to connect the Moroso oil pump and pickup to the rear main cap. Taylor also used a new ARP oil-pump driveshaft.

Fel-Pro offers this very slick one-piece oil-pan gasket that makes sealing the pan a breeze. It literally drops in place without sealant, and your oil-pan leaks are a thing of the past. Each bolt hole also incorporates steel-limiter bushings to prevent overtightening. This is one slick piece. Use this gasket once and you'll never go back to a four-piece standard oil-pan gasket.

engines, which led to a pair of healthy Xtreme Energy mechanical-roller cams. The valvetrain included Comp's mechanical-roller tappets, the recommended dual valvesprings, retainers, keepers, a special ceramic-tipped mechanical fuel pump pushrod, and proper-length pushrods. These are combined with the remainder of the valvetrain that includes a Comp timing set, a thrust button, and a two-piece timing cover that allows us to move the cam if necessary. To button up the valvetrain, we topped it all off with a set of Comp's bulletproof Hi-Tech stainless 1.52:1 roller rockers.

Any cylinder head would have worked in this situation, but we decided to try and keep a lid on the price of power by going with a set of Trick Flow Specialties' new 23-degree aluminum small-block cylinder heads. These are the same heads we used on the Goodwrench Quest 350 that made 416 hp.

The TFS heads are offered in three different configurations depending upon valvespring diameter. Beyond that, the heads feature a 195cc intake port, a 64cc combustion chamber, and 2.02/1.60-inch stainless steel valves. We chose these heads mainly because

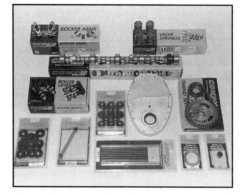 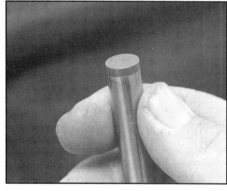

Comp Cams supplied the entire valvetrain system (left) that included the roller cam, mechanical-roller lifters, timing-chain set, springs, retainers, and keepers. Because this is a roller cam, we also needed an adjustable roller-cam button and a special ceramic-tipped mechanical–fuel-pump pushrod (right) to protect the steel roller cam. Blanketing all this is a Comp Cams two-piece timing-chain cover to make cam swaps a snap.

they offer an outstanding dollar-per-horsepower potential and would complement the selection of the Comp Cams roller cam.

Heads can't help make horsepower if you don't match them up with a decent induction system. This test called for an Edelbrock Victor Jr. single-plane intake mainly because we wanted to create the least amount of inlet restriction possible. Combining the potential of both the cylinder heads and the roller cam, this made the decision to go with the Victor Jr. easy. We also matched the intake with an aluminum Edelbrock water pump to keep everything cool.

Obviously, the ignition system needed to be up to the task, so we added an MSD billet distributor along with MSD's new Digital 6 ignition box and Super Conductor plug wires. Keeping all these nice parts together and leak-free fell to a full set of ARP fasteners including head bolts, main studs, oil-pump drives, crank bolts,

Battle of the Small-Block Strokers

Taylor used a Tavia camshaft handle to install the roller cam in the short-block. Do not use thick moly-lube on a roller cam. Instead use engine oil or engine-assembly lube (the thicker red oil) to coat the cam and roller lifters.

The new kid on the Comp Cams block are these Xtreme mechanical-roller lifters. The tiny slot aims pressurized oil directly at the roller-lifter interface with the cam to ensure proper lubrication even at very low engine speeds. This prevents galling of the tappet and cam as well as ensures sufficient lubrication to the roller bearings.

Since roller cams tend to "walk" fore and aft, all roller cams use a thrust washer (left photo) between the cam gear and the block to prevent block wear. The roller-thrust button (arrow, right photo) limits camshaft endplay by the bearing against the timing-chain cover. Roller cams are limited to minimal endplay to prevent cam walk that can cause retarded ignition timing due to the helical cut of the distributor drivegear. We used ARP cam bolts to keep everything snug.

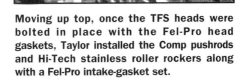 

The Summit 23-degree head combines 2.02/1.60-inch valves and a 64cc chamber to create excellent airflow.

Moving up top, once the TFS heads were bolted in place with the Fel-Pro head gaskets, Taylor installed the Comp pushrods and Hi-Tech stainless roller rockers along with a Fel-Pro intake-gasket set.

and a full dress-up kit for both engines. Other essential pieces used to complete the construction of both engines included ATI's outstanding torsional dampers, Fel-Pro gaskets throughout both engines, Plasti-Kote Electric Orange paint, and a CSI billet swivel water neck.

If you look closely at the parts list for both engines, there are a few subtle differences worth noting. The two-bolt main 400 block required more strength, so we added Gellner steel four-bolt main caps in the center three caps secured with ARP main studs. This required the caps and block be align-honed to establish a proper size and placement of the main-bearing bores. When the 400's main bores are align-honed, this enlarges the rear-main-seal bore diameter, requiring a specific Fel-Pro 400 rear main seal. The 400 block for the 377ci engine also required 400-style head gaskets. Since this is a street engine, the head gaskets required vent holes between the cylinders to prevent the formation of steam pockets in these areas. This is a critical step for any 400-based street

Small-Block Chevy Engine Buildups

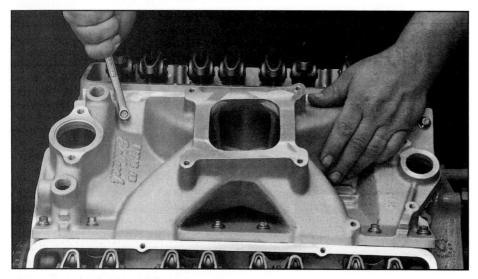

Next came the Edelbrock Victor Jr. single-plane intake manifold. Since we were looking for overall best power, we didn't want to limit airflow with a dual-plane intake. It's possible that an Edelbrock Performer RPM would have improved torque on both engines, but peak horsepower would probably have suffered.

You can't make good horsepower without a decent carburetor, so we bolted on a Barry Grant 750-cfm double-pumper Speed Demon carburetor. The large glass-sight plugs make setting the float level easy, and the billet construction ensures an excellent fuel curve.

To finish off both Donnie and Marie, Taylor added ATI Super Dampers that are degreed friction dampers which do a superb job of damping crankshaft vibration at all engine speeds. That slick timing indicator is from Tavia, while the crank bolt is an ARP item.

engine. Taylor then drilled matching steam holes in the cylinder heads as well.

But that's enough talk. Pictures tell the story much better than just mere words, so we'll let the photos tell the rest of the story.

DYNO TESTING
The 377

The first engine to roll into the dyno cell was Donnie, our rock-'n'-roll 377ci engine. After a sufficient warm-up and a quick run-through of the timing and jetting curves, we were ready to twist this motor up. The carburetor we were using was a GTP Speed Demon 750-cfm double-pumper. After a series of jetting exercises, we used a set of Comp Cams MaxJets to come up with 84 primary and 96 secondary jetting. Timing finally peaked with a best power at 36 degrees of spark lead.

This first series of tests was run with a set of Hedman Elite–series 1 3/4-inch standard-length headers followed up by a 2 1/2-inch exhaust system muffled with a pair of Borla XR1 stainless steel mufflers. Later, we would also test both engines using a shorter set of intermediate-length headers to see if either engine preferred a shorter-length header.

After several tests, the 377 ended up with a very well-behaved torque curve with peak torque arriving at 4,900 rpm at 450 lb-ft of torque, while horsepower peaked at 6,100 rpm with 454 hp. This created a peak-to-peak powerband of 1,200 rpm, which is fairly typical of most small-blocks. The interesting part was this fairly small 195cc Summit intake-port aluminum cylinder head created enough torque that this supposedly peaky bore-stroke combination was making 412 lb-ft even at 3,000 rpm.

Generally, most 9.0:1-compression street engines make around 1.1 hp/ci. This is a relatively good number and the place where most well-built street engines find themselves. A little quick math revealed that Donnie produced 1.20 hp/ci and a very credible 1.19 lb-ft per inch. These are very good numbers, which made us wonder if the 383

Battle of the Small-Block Strokers

We'd show you photos of both engines on the dyno, but there's really no reason to because they look exactly the same! This is the 377 (at least we think it's the 377) on Duttweiler's dyno prepared to pump 454 hp through the absorber. Those are Zoop's billet-aluminum pulleys along with Hedman 1 3/4-inch headers.

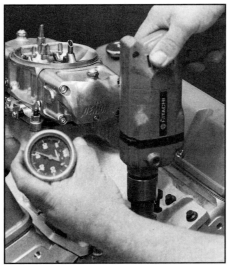

Before we fire up any new engine, we always pre-oil the engine using an ARP pre-oiler to spin the oil pump using a 1/2-inch drill motor. This reduces start-up wear on the bearings and makes for a happy engine.

Part of the buildup included torquing the ARP 7/16-inch studs into the Summit 23-degree aluminum heads.

motor could match them. We cooled Donnie's heels and decided to bolt on Marie and see what she could do.

The 383

With the 383 now bolted to the dyno pump, we duplicated our break-in procedures and also duplicated both the timing and jetting numbers to see if the 383 wanted anything different. Because the displacement is so similar between the two engines and all the other components are the same, we quickly found that the 377's prior best tune-up was exactly what the 383 wanted as well, so nothing was changed in this department. With the oil temperature equal to the previous 377ci engine test, Duttweiler pulled the handle and let the 383 become the torque of the town.

We decided to look at the 383's power curve independently rather than compare it to the 377 right out of the box. Immediately, it was obvious that the 383 also delivered a strong torque curve just as we anticipated it would. The 383 took off with the torque twisting up 458 lb-ft of torque at 4,600, while horsepower came stormin' in at 450 ponies at 6,000 rpm. This put the 383 at 1.17 hp/ci while equaling the 377's torque per cubic inch at 1.19.

Looking back at this test now, it would have been fun to measure torque between 2,000 and 3,000 rpm because it looks like the 383 would deliver well over 400 lb-ft of torque even as low as 2,200 to 2,300 rpm. Advancing the camshaft a couple of degrees could have enhanced this with longer-header collectors. But even with the 383 in its current configuration, this is an excellent combination of torque and horsepower.

POWER COMPARO

Now it was time to compare the two power curves. A quick glance at the "Difference" column on the graph illustrates that the theory tends to hold true. The 377ci engine delivered a horsepower advantage based on its larger bore, while the 383 pumped up the torque curve below 5,000 rpm. But which one is better?

Initially, horsepower pundits may point to the 377's 16 hp advantage at 6,300 rpm over the 383. While on the other side of the rpm curve, torque talkers will point to the 383's 18-lb-ft advantage at both 3,300 and 3,500 rpm. But the reality is you really should look at averages over the entire power curve rather than individual data points.

Looking first at average torque below 5,000 rpm, it's clear that the 383 eats the 377's lunch. The 383 scores big-time points here as a powerful street engine since it offers an average of 13-lb-ft more torque at every point between 3,000 and 4,900 rpm. But then if you look at horsepower from 5,000 to 6,700 rpm, the 377 starts to really shine. Here, the differential is only an average of less than 7 hp.

It's also worthwhile to look at the power numbers beyond peak horsepower. The 383 drops 17 hp at

Small-Block Chevy Engine Buildups

Final assembly involved setting the lash on the Comp Cams Xtreme Energy XR 280R mechanical-roller (0.016/0.018-inch intake and exhaust) using the Comp Hi-Tech stainless 1.5:1 roller rockers.

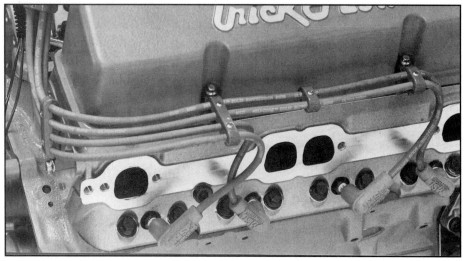

The power key to both engines revolves around these Summit aluminum heads. The intake ports measure 195 cc, which is a great size that creates a great cross-breeding of torque, horsepower, and street manners. We added a set of Trick Flow cast-aluminum valve covers as well as a set of MSD plug wires held in place with Made For You plug-wire looms.

6,500 from its 6,000-rpm 450 hp peak, while the 377 carries its power much better past peak, losing only 6 hp at the same 6,500-rpm point. This means you'd be able to pick up e.t. on the dragstrip by shifting the 377 at perhaps a couple hundred rpm higher than the 383 engine. This is especially important if this 377 is used with a Turbo 350 or 400 automatic. The rpm drop between gears with a three-speed automatic usually drops the rpm below the engine's torque peak. By raising the shift point to 7,000 or 7,200, the engine will recover at a higher rpm closer to the torque peak, which will improve the e.t.

Clearly, the 383 makes more torque while the 377 makes more horsepower. So which one is the winner?

The answer we came up with is that from an overall power standpoint, the 422.8 lb-ft of average torque generated by the 383 was a little over 4 lb-ft more than the 377's 418.5-lb-ft average. Plus, the added torque from 3,000 to 4,600 just makes the 383 a much more fun engine for a street car than the 377.

This tends to hold true for a dragstrip application as well. We simulated both engines in a '68 Camaro with the Racing Systems Analysis Quarter Pro program using a 3,400-pound Camaro with 3.55 gears, a Turbo 350 trans with a 3,000-stall converter, and 8.5-inch slicks that were 26 inches tall. The 377 package ran a best of 11.69/117.6 mph, while the 383 blasted to a 0.05-second quicker 11.64/117.5 effort. We also tested the higher shift-point idea with both engines. By increasing the shift point from 6,500 to 7,000 rpm, both engines picked up e.t. with the 377 improving slightly more at 0.05 second and 0.70 mph.

CONCLUSION

As you can see, both engines are very close in power. It appears the 383 would make the better street engine, while the 377 would make a good road-race motor if you were willing to wind this engine a little tighter. With better cylinder heads, the 377 might end up with a more significant increase in horsepower, but both Donnie and Marie are winners in our book.

Battle of the Small-Block Strokers

Terry Zupan came by to help put Marie on the dyno and topped off the engine with an Edelbrock Victor Jr. single-plane intake and a 750-cfm Race Demon carburetor.

We experimented with several jetting combinations on the billet-aluminum metering plates in the Race Demon carburetor before optimizing power with a set of 84 primary and 92 secondary Comp Cams MaxJets. These MaxJets are numbered differently than Holleys. The conversion is roughly 84 MaxJet = 78 Holley and 96 MaxJet = 86 Holley.

We also used an Edelbrock aluminum water pump, an ATI SuperDamper harmonic damper, and a Carter mechanical fuel pump to complete engine externals. That's a Comp Cams two-piece timing-chain cover, which makes changing cams and cam timing much easier.

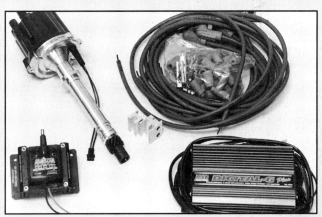

The ignition side of both engines ensured plenty of spark with a new MSD Digital-6 spark box and matching coil. To ensure all that spark energy made it to the spark plugs, Ed Taylor custom-built a set of wires using MSD's Super Conductor 8.5mm wires.

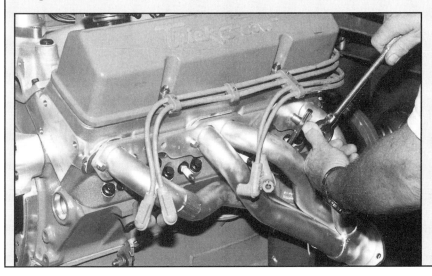

We also tried a set of Hedman intermediate headers on both engines to see if one or both would respond to the shorter primary-pipe-diameter headers.

Small-Block Chevy Engine Buildups

PARTS LIST

COMPONENT	MANUFACTURER	PART NUMBER
Block 383ci, 350, four-bolt	GM	Used
377ci, 400, two-bolt	GM	Used*
Four-bolt, 400 main caps	Gellner Engineering	SBC-400S*
Crank, 383ci	Trick Flow	TFS-35375300
377ci	Trick Flow	TFS-35605C60*
Connecting rods	Trick Flow	TFS 36600500
Pistons 383ci, dished	SRP	138103
377ci, dished	JE	Custom-order*
Rings, 383ci, file-fit	Speed-Pro	R-9771-35
377ci, file-fit	Speed-Pro	R-9346-35*
Bearings, main, 383ci	AE Clevite	MS 909H
377ci	AE Clevite	MS1038H*
Bearings, rod	AE Clevite	CB663H
Bearings, cam	AE Clevite	SH290S
Head gaskets, 383ci	Fel-Pro	1003
Head gaskets, 377ci	Fel-Pro	1004*
Rear main seal, 377ci	Fel-Pro	2909*
Rear main seal, 383ci	Fel-Pro	2900
Race set	Fel-Pro	2702
Pan-gasket set	Fel-Pro	1821
Valve-cover set	Fel-Pro	1628
Camshaft, XR-280-10, roller	Comp Cams	12-771-8
Lifters, roller	Comp Cams	888-16
Pushrods	Comp Cams	7693-16
Rocker arms, 1.52:1	Comp Cams	1304-16
Timing chain	Comp Cams	3100
Pushrod, fuel pump for roller	Comp Cams	4608
Valvesprings, roller-style	Comp Cams	978-16
Retainers	Comp Cams	740-16
Cam button	Comp Cams	200
Thrust washer	Comp Cams	201
Timing cover, two-piece	Comp Cams	210
Cylinder heads	Trick Flow	TFS-30400003
Valve covers	Trick Flow	TFS-31500802
Damper	ATI	917780
Oil pan	Moroso	20191
Oil pump and pickup	Moroso	22135
Intake manifold	Edelbrock	2975
Water pump	Edelbrock	8811
Carb, 750-cfm Speed Demon	GPT	1402010
Distributor	MSD	85551
Ignition box, Digital 6	MSD	6520
Plug wires	MSD	3120
Wire separators	Made For You	50-156
Head bolts	ARP	134-3601
Dress-up bolt kit	ARP	534-9801
Crank-bolt kit	ARP	134-2503
Oil-pump stud	ARP	230-7003
Main studs, 383ci	ARP	134-5601
Main studs, splayed, 377ci	ARP	234-5602*
Oil-pump drive	ARP	134-7901
Timing pointer	Tavia	02344
Fuel pump	Carter	M4891
Spark plugs	AC	FR2LS
Water neck, billet swivel	CSI	912-B
Block paint	Plasti-Kote	200

*Denotes specific parts for the 377ci engine that differ from the 383ci

HEADER COMPARISON

In this test, we compared the Hedman intermediate headers to the long-tube headers for both engines. In both cases, you can see that the shorter length intermediate headers suffered a substantial torque loss while delivering only a mild horsepower increase at the top end. For a street-driven vehicle, the smart move here would be to run the longer-tube headers since they make substantially more torque that will accelerate the car harder in the mid-range. For this test, we only looked at the torque differences. A torque increase at high rpm equals more horsepower.

RPM	TEST 3 377ci TQ	TEST 4 383ci TQ
3,000	-19	-27
3,100	-22	-25
3,200	-18	-27
3,300	-15	-27
3,400	-22	-25
3,500	-22	-30
3,600	-22	-28
3,700	-18	-28
3,800	-20	-27
3,900	-17	-31
4,000	-11	-26
4,100	-6	-25
4,200	-7	-21
4,300	-5	-18
4,400	-1	-16
4,500	+2	-3
4,600	0	-6
4,700	-3	-2
4,800	-3	-4
4,900	-1	-3
5,000	+4	-1
5,100	+4	0
5,200	+3	+2
5,300	+4	+2
5,400	+4	+1
5,500	+4	+6
5,600	+2	+8
5,700	+7	+4
5,800	+7	+10
5,900	+7	+6
6,000	+9	+9
6,100	+8	+11
6,200	+8	+9
6,300	+7	+13
6,400	+14	+14
6,500	+10	+12
6,600	+12	+4
6,700	+12	+8
6,800	+11	+3

Battle of the Small-Block Strokers

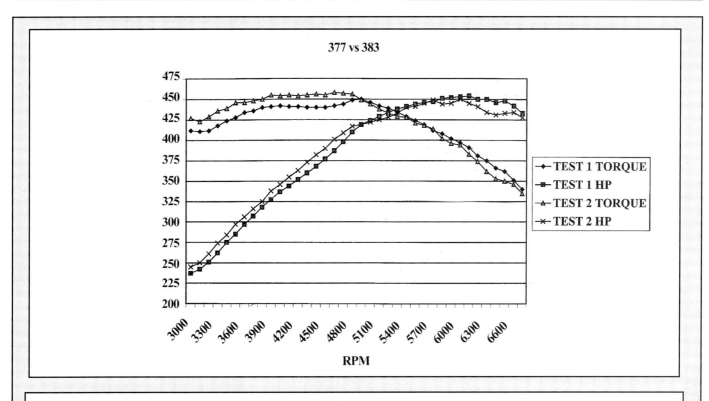

DYNO TESTING

TEST 1: 377ci engine with long-tube, 1 3/4-inch Hedman headers, Race Demon 750 carb with MaxJet 84 primary and 96 secondary jets, and Borla XR-1 mufflers with 2 1/2-inch full exhaust system

TEST 2: 383ci engine with the exact same configuration of external components as Test 1

The Difference column shows the 377 versus the 383. Note that the 383 makes more torque at the lower end of the rpm scale, while the 377 takes over above 5,000 rpm.

	TEST 1 (377ci)			TEST 2 (383ci)			DIFFERENCE	
RPM	TQ	HP	BSFC	TQ	HP	BSFC	TQ	HP
3,000	412	237	0.406	427	245	0.410	-15	-8
3,100	411	242	0.410	423	250	0.410	-12	-8
3,200	412	251	0.406	429	261	0.403	-17	-10
3,300	418	262	0.400	436	274	0.400	-18	-12
3,400	424	275	0.393	439	284	0.400	-15	-9
3,500	428	285	0.393	446	297	0.390	-18	-12
3,600	434	297	0.395	446	306	0.390	-12	-3
3,700	436	307	0.397	448	316	0.390	-12	-9
3,800	440	318	0.397	450	325	0.390	-10	-7
3,900	441	328	0.397	455	338	0.395	-14	-10
4,000	442	337	0.398	454	346	0.400	-12	-9
4,100	441	344	0.401	455	355	0.390	-14	-11
4,200	441	352	0.400	454	363	0.390	-13	-11
4,300	440	360	0.400	455	373	0.390	-15	-13
4,400	440	368	0.399	456	382	0.380	-16	-14
4,500	440	377	0.399	455	390	0.380	-15	-13
4,600	442	387	0.397	458*	401	0.390	-16	-14

Small-Block Chevy Engine Buildups

RPM	TEST 1 (377ci)			TEST 2 (383ci)			DIFFERENCE	
	TQ	HP	BSFC	TQ	HP	BSFC	TQ	HP
4,700	444	398	0.396	457	409	0.390	-13	-11
4,800	449	410	0.395	456	417	0.390	-7	-7
4,900	450*	419	0.399	449	419	0.400	+1	0
5,000	446	424	0.403	444	422	0.410	+2	+2
5,100	442	429	0.408	438	425	0.410	+4	+4
5,200	439	434	0.413	433	428	0.420	+6	+6
5,300	434	438	0.420	429	433	0.420	+5	+5
5,400	429	441	0.420	428	440	0.420	+1	+1
5,500	424	444	0.422	421	441	0.430	+3	+3
5,600	419	446	0.420	418	445	0.420	+1	+1
5,700	412	447	0.430	414	449	0.420	-2	-2
5,800	408	451	0.430	402	444	0.434	+6	+7
5,900	402	452	0.440	396	445	0.450	+6	+7
6,000	397	453	0.440	394	450*	0.450	+3	+3
6,100	391	454*	0.450	383	445	0.460	+8	+9
6,200	381	450	0.457	374	441	0.480	+7	+9
6,300	375	450	0.460	362	434	0.500	+13	+16
6,400	366	446	0.460	353	431	0.520	+13	+13
6,500	362	448	0.460	350	433	0.521	+12	+15
6,600	351	442	0.470	346	434	0.520	+5	+8
6,700	340	433	0.490	335	428	0.520	+5	+5

*Denotes peak-power numbers

Power per Cubic Inch
377 ci - 454 hp/450 torque = 1.20 hp/ci and 1.19 lb-ft of torque/ci
383 ci - 450 hp/458 torque = 1.17 hp/ci and 1.19 lb-ft of torque/ci

Average Power

Displacement	377.0	383.0
Average torque below 5,000	434.2	447.4
Average horsepower 5,000 and above	443.4	437.1
Average overall torque	418.5	422.8

Flow to Go

Flow Testing Small-Block Chevy Cylinder Heads
By Jeff Smith
Photography by Jeff Smith, John Baechtel and Mike Petralia

17

Horsepower is where it's at, baby. And the best place to find it is with good cylinder heads. A great way to evaluate a cylinder head is to measure how much air you can push through it. That's what a flow bench does. Since the small-block is the most popular engine in Chevrolet's arsenal, we decided to take on the daunting task of flow-bench testing every significant small-block Chevy street cylinder head, both stock and aftermarket.

HEAD BASICS

Small-block Chevy cylinder heads can be categorized in several different ways. One of the easiest is to group them by intake-port volume. For the purposes of this chapter, we will organize these heads into three categories: Category 1, 179 cc and smaller; Category 2, 180 cc to 199 cc; and Category 3, 200 cc to 220 cc. Unmolested production heads generally fall in Category 1, but not always. For this test, we've evaluated most of the popular heads. For the earlier heads, we'll refer to them by the last three digits of their casting number, as in 462 heads, which were early castings that had a 64cc combustion-chamber volume and were used on 327 and 302 engines. The later heads will be referred to by their application, such as the aluminum L98 TPI castings, the LT1 and LT4 5.7L engines, as well as the cast-iron Vortec head and the new LS1.

Flow-testing 41 cylinder heads is not an easy task. John Baechtel performed all of the cylinder-head testing on a SuperFlow 600 flow bench equipped with SuperFlow's latest computerized FlowCom digital data-capture feature. This not only allows quicker testing but helps eliminate test variables.

While we have separated these heads by port volume, greater size doesn't guarantee increased airflow. The advantage to midsize heads is that they combine increased airflow without sacrificing velocity, which tends to promote strong torque. One way to look at cylinder-head flow is to include velocity when evaluating heads. In other words, an intake port the size of a tennis ball will probably flow tons of air but at a snail-like velocity. Conversely, smaller port volumes will suffer from flow restrictions but offer fantastic velocity. Velocity, then, can be equated with good torque and excellent throttle response.

If there is a magical combination, it would have to be a port with great airflow and outstanding velocity. This is especially true for intake ports. A good example of an extra-large intake port that didn't work would be the '69–'70 Ford Boss 302 engine. The ports on that engine are huge. Those engines made decent horsepower but were infamous for their lack of

Small-Block Chevy Engine Buildups

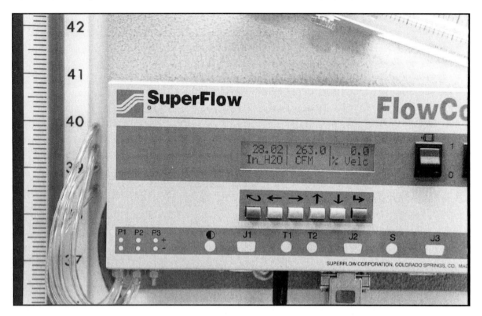

The Superflow FlowCom digital airflow-measurement system is a retro-fit system that can be applied to any SuperFlow 600 bench. An internal motor controls and stabilizes the test pressure and allows more accurate flow-data recovery. At a later date, Baechtel will add SuperFlow's WinFlow software program, which can create custom graphs, charts, and other ways of evaluating data.

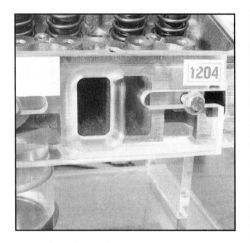 

Prior experience has shown that a radiused inlet must be used when testing intake ports. Brzezinski makes several plastic radiused inlets for the small-block Chevy based on Fel-Pro intake gasket sizes (left). But for several heads, including the LT1, LT4, and the new LS1 port (right), no dedicated radiused inlet was available. For these heads, we used modeler's clay to shape a similar radius. The LS1's intake-port shape is designed to allow the electronic fuel injector to aim fuel directly into the cylinder with the valve open.

torque. This is also due to cam timing but mainly to the expansive intake ports. That's why pure air flow numbers can be misleading. You must also consider the application for which the cylinder will be used. For example, it would be folly to bolt a huge 220cc intake port head on a daily-driven 283ci street car.

HEAD GAMES

There's much more to cylinder-head selection than just choosing the head with the best flow numbers. In fact, as we've explained, large intake-port volumes with high peak-flow numbers may not be the right choice at all. There are actually several criteria for selecting the best overall cylinder heads for your application. The best way to look at cylinder-head flow is to closely examine and compare the entire flow curve from low lift to max lift.

If you must look at only one point, most airflow experts prefer to look at mid-lift flow. For street engines, airflow at 0.400 inch is a good place to evaluate all cylinder heads. The main reason for this is that maximum lift flow at 0.550- or 0.600-inch valve lift is at the ragged edge of most streetable valvetrains. More importantly, the valve is only at max lift once in the entire valve-lift curve, while the valve achieves mid- and low-lift numbers on both the opening and closing sides of the valve-lift curve.

Cylinder-head selection also means evaluating exhaust-flow numbers as well. A cylinder head with killer intake numbers but poor exhaust flow is not going to perform as well as a head with a slightly weaker intake but an excellent exhaust port. It is possible to prop up a weak exhaust port with longer exhaust duration and more lift, but generally that engine will not perform as well as an engine with a strong exhaust port and a more ideal camshaft. One way to evaluate the relative strength of an exhaust port is by comparing the exhaust flow to the intake flow as a percentage at the same valve lift. This is generally referred to as the exhaust-to-intake (E/I) relationship. For example, let's say the intake port flows 250 cfm at 0.400-inch valve lift and the exhaust port flows 187.5 cfm. Exhaust-port flow divided by intake-port flow will express this relationship as a percentage. In this case, the exhaust flows 75 percent of the intake at the same valve lift.

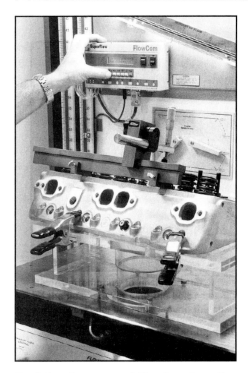

Each head was carefully placed on the SuperFlow bench and then tested from 0.050- to 0.600-inch valve lift. In some cases, the heads were tested twice to ensure that the numbers were accurate. All tests were corrected for temperature, a procedure few flow-bench testers bother to go through when flow-testing.

Combustion chambers were also checked to establish combustion-chamber size. The trend with later heads is certainly a movement toward smaller, tighter combustion chambers that enhance combustion efficiency.

Conventional wisdom holds that anything more than a 75 percent E/I relationship is considered good. Eighty percent is even better. But anything more than this might be suspect. In other words, a great E/I may point to a weak intake port rather than to a strong exhaust port.

PIPE DREAMS

One of the most recent changes in cylinder-head flow-testing is the use of an exhaust pipe. The idea is to better simulate the way the exhaust port flows air. Our testing revealed that there is a significant difference in exhaust-port flow when the same port is flowed with and without a pipe. Because there are proponents of both ways of testing exhaust ports, we decided to test using both configurations. Higher lift numbers tended to reflect the greatest change in flow between using the pipe and not using it. Increases were around 4 to 7 percent with the addition of the flow pipe.

Valves play an important part in cylinder-head flow. In most, but not all, cases we tested each head using the valves supplied with the cylinder head. One easy way to enhance low- and midlift flow is to perform a 30-degree back cut on the intake valve (and sometimes on the exhaust valve). This radiuses the flow around the valve angle much as a three-angle valve job does on the seat. Not all valves used in this test had back cuts on the intake.

HEAD SELECTION

Once you're armed with all the information in the test, the key is proper selection of the right cylinder head for your combination. As is the case with most things high-performance, bigger is not necessarily better. For mild street engines of 350 ci or less that rarely see the high side of 5,500 rpm, there's little reason to select a cylinder head with a port volume of more than 200 cc. Engines in this category respond better to a smaller intake-port volume that will contribute to excellent low- and midrange torque. This also contributes to excellent throttle response and even decent fuel economy when teamed with a mild camshaft, a dual-plane intake manifold, and a carburetor with small primary throttle bores, such as a Quadrajet- or Carter-style carburetor.

Engines of larger displacement, in the range of 383 to 406 ci or heftier, can take advantage of a larger intake port, depending upon how the engine is intended to be used. For example, a

Small-Block Chevy Engine Buildups

406 small-block that will end up in a four-wheel-drive Chevy truck would be better served with an intake port of between 180 and 190 cc, since this smaller port volume will again promote excellent torque from this larger-displacement engine. For an engine intended for higher-rpm horsepower, a larger port volume will tend to promote greater horsepower when combined with a longer duration camshaft. For example, a 210cc intake-port cylinder head is rather large for a 383ci small-block, but if you plan to spin this motor to 6,500 or even 7,000 rpm, the large intake-port head might contribute to a significant power increase.

As you can see from this rather brief overview, the business of selecting a good cylinder head can be downright confusing. Port volume, port flow, E/I percentages, cost, weight, chamber size, valve size, and probably at least another half-dozen or so variables contribute to making a cylinder-head choice rather difficult. On the other hand, spending a little more money for an aftermarket cylinder head is usually a very good idea. Airflow is the key to making horsepower, and all of the aftermarket heads we tested offer improved airflow over stock, 25- to 30-year-old castings. To paraphrase engine builder John Lingenfelter,

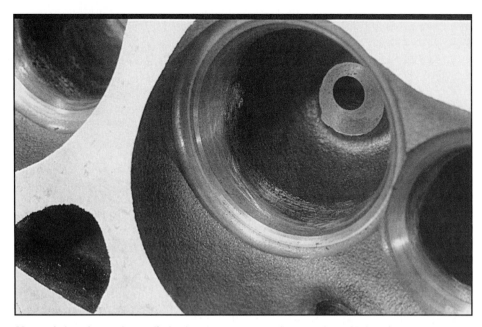

Many of the aftermarket cylinder heads were treated to hand- or CNC-radius work directly below the valve-seat area. This kind of blending can have a dramatic and positive effect on port flow. The greater the sharp edge underneath the machined valve seat, the greater the effect of this blend work on airflow. Some of the heads tested did not have this work, which certainly affects their airflow in this test. Most aftermarket cylinder-head companies offer this at an additional cost.

"The world's best cam combined with a poor set of heads will produce an engine that's a dog. But bolt on a set of great heads even with a poor cam, and that engine will still make great power."

CATEGORY 1 SMALL-BLOCK HEADS: UNDER 180 CC

GM 882 IRON
The smallest of the GM heads we tested at only 151 cc, the 882 turned out to be the best-flowing of the early iron heads. This head was tested with the 1.94/1.50 valves. But be aware that adding 2.02/1.60-inch valves without blending in the short side radius usually results in lost airflow. Previous testing with other heads, including the 441, revealed the 441 to flow better than the 882. If you are into these heads, the best way to find a good one is to flow-test several. But in most cases, that's impractical. Port: 151 cc. Chamber: 76 cc. Valves Sizes: 1.94/1.50. I Flow @ 0.400: 203 cfm

Valve Lift	Intake	Exhaust Open	Exhaust w/pipe
0.050	39	34	34
0.100	70	58	59
0.200	125	108	109
0.300	175	135	136
0.400	204	141	143
0.500	205	142	144
0.600	206	142	145
E/I		73%	70%

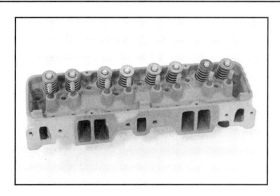

GM 441 IRON
The 441 has a reputation as a good-flowing head among the production 76cc chamber heads. In this particular case, it lost out to the 882 head, but overall it's a decent head that responds well to pocket port work and larger 2.02/1.94 valves. Its small 155cc port volume creates great torque and excellent driveability. While produced in the thousands, this head will become increasingly hard to find, much like the 462 head is now. Port: 155 cc. Chamber: 76 cc. Valve Sizes: 1.94/1.50. I Flow @ 0.400: 194 cfm.

Valve Lift	Intake	Exhaust Open	Exhaust w/pipe
0.050	30	22	20
0.100	55	45	45
0.200	110	80	83
0.300	160	112	118
0.400	194	127	133
0.500	201	136	146
0.600	203	139	148
E/I		65%	68%

GM 462 IRON
The 462 is one of the small-chamber heads used on 327s. The higher-horsepower applications used larger valves, but the ports were the same. The 462 head is representative of the small-chamber, 64cc

heads. These heads are becoming very hard to find since they were discontinued at the end of 1968. If these heads are used on engines of 350 ci or larger, dished pistons are required since the small chamber raises the compression. Pocket porting and larger valves can be added to make these heads work well, but the investment of money and time may not be the best use of your resources, especially when you consider that the World Products S/R Torquer heads are not only brand-new castings but are inexpensive and flow better. Port: 156 cc. Chamber: 62 cc. Valve Sizes: 1.94/1.50. I Flow @ 0.400: 198 cfm

Valve Lift	Intake	Exhaust Open	Exhaust w/pipe
0.050	18	27	27
0.100	50	54	54
0.200	119	88	91
0.300	167	116	121
0.400	198	128	134
0.500	212	134	139
0.600	218	137	142
E/I		64	67

Small-Block Chevy Engine Buildups

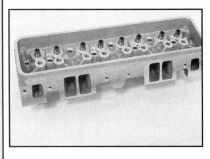

GM L98 ALUMINUM

This is the aluminum head used on the Corvette TPI engines from the mid-'80s up until Chevrolet converted to the LT1 engine in 1992. While this L98 head has a decent reputation on the street, our flow-testing shows that the 882 iron head outflows it across the board. Couple this with the L98 head's small 58cc chamber, and it's tough to choose this head as a performance street head. The only thing it has going for it is that it's aluminum. With the steel valve inserts, the largest valve you can stuff in this head is a 2.00/1.55-inch package. Unless you find these heads really cheap, there are several aftermarket heads with better flow that would make better choices. Port: 163 cc. Chamber: 58 cc. Valve Sizes: 1.94/1.50. I Flow @ 0.400: 186 cfm.

Valve Lift	Intake	Exhaust Open	Exhaust w/pipe
0.050	29	27	27
0.100	56	49	52
0.200	120	96	104
0.300	160	123	133
0.400	186	140	150
0.500	196	155	170
0.600	199	159	175
E/I		75%	80%

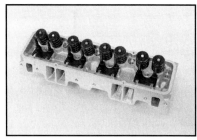

EDELBROCK PERFORMER

The Edelbrock Performer is intended as a great overall performance cylinder head. Its smaller 166 cc intake port volume makes for great torque when combined with its great flow numbers. The E/I flow relationship is good, and when viewed as a complete package, the Performer heads can create a great street package. We have personal experience with these heads on a 9:1 355 making more than 420 hp at 6,000 rpm with a Comp Cams DEH 275 cam and a Performer RPM intake. It also made almost 440 lb-ft of torque at 4,250 rpm. This is excellent power with outstanding torque. Don't be fooled by the conservative flow numbers. This head works. Port: 166 cc. Chamber: 70 cc. Valve Sizes: 2.02/1.60. I Flow @ 0.400: 216 cfm

Valve Lift	Intake	Exhaust Open	Exhaust w/pipe
0.050	32	20	20
0.100	64	44	44
0.200	121	95	96
0.300	174	134	134
0.400	216	159	159
0.500	235	177	177
0.600	236	187	187
E/I		73%	73%

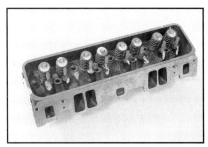

GM VORTEC IRON 855

Here is perhaps the sleeper head among GM's factory castings. According to our sources, this head was designed after the LT1 aluminum head and offers better flow over the LT1, but only on the intake side, which we confirmed with our flow-testing. This Vortec truck head is also the same head used for the iron LT1 engine used in the Impala SS, just with a different intake-manifold-bolt pattern. Compare the Vortec flow numbers to the LT1, and you'll see the Vortec moves some air, especially in the midlift area. The small chamber will require a dished piston with an engine of 350 or larger to keep compression under control for pump gas, but this is a good head for situations in which iron heads are required. This Vortec head even outflows the now-antiquated iron Bow Tie head. Port: 170 cc. Chamber: 64 cc. Valve Sizes: 1.94/1.50 I Flow @ 0.400: 227 cfm.

Valve Lift	Intake	Exhaust Open	Exhaust w/pipe
0.050	40	25	25
0.100	70	48	49
0.200	139	101	105
0.300	190	129	137
0.400	227	140	151
0.500	239	147	160
0.600	229	151	162
E/I		62%	66%

GM LT1 372 ALUMINUM

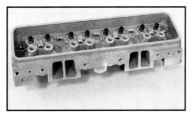

The LT1 was the next-generation head developed for the small-block after the L98 TPI engine. The LT1 head uses a unique reverse-cooling system, which means it cannot be swapped onto older small-blocks. As you can see from the 0.400-inch lift-flow number, this is a great head right out of the box at 213 cfm. Several companies, including Lingenfelter, offer ported versions of this head that really rock 'n' roll. Larger 2.00/1.55-inch valves can also be added to improve flow slightly along with some short-side radius blending. With this kind of airflow, you can see why these engines make great power. Port: 170 cc. Chamber: 58 cc. Valve Sizes: 1.94/1.50. I Flow @ 0.400: 213 cfm

Valve Lift	Intake	Exhaust Open	Exhaust w/pipe
0.050	42	28	30
0.100	72	52	53
0.200	129	100	101
0.300	180	130	136
0.400	213	142	155
0.500	214	149	165
0.600	215	154	171
E/I		67%	73%

WORLD PRODUCTS S/R TORQUER

The World Products torquer head is the larger valve version of World's Stock/Replacement (S/R) casting, which, as its name implies, is the straight replacement for the production iron heads. The Torquer offers great airflow for a low price with virtually identical flow at 0.400-inch valve lift to the 882 head but is better by 15 cfm at 0.500-inch valve lift. With machining costs to rebuild stock heads increasing all the time, this S/R Torquer is a great alternative to rebuilding stock heads, with improved flow being an additional reason to decide on these heads over stockers. Port: 170 cc. Chamber: 67 cc. Valve Sizes: 2.02/1.60. I Flow @ 0.400: 205 cfm

Valve Lift	Intake	Exhaust Open	Exhaust w/pipe
0.050	36	27	28
0.100	68	55	55
0.200	128	100	104
0.300	179	133	140
0.400	205	145	156
0.500	221	153	166
0.600	236	155	170
E/I		71%	76%

DART IRON EAGLE S/S

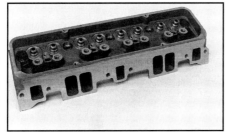

Dart has recently filled out its inventory of small block Chevy iron heads with everything from this basic stock replacement 165cc intake port head called the Street/Strip all the way to 230cc Iron Eagle monsters. This smallest head outflows a stock Chevy 441 on the intake side and exhaust except at the top of the curve. The Dart even keeps up with an old 462 style casting, so this would make a great head to replace those aging original castings. The Iron Eagle S/S 165 only comes in a straight-plug configuration with a 76cc combustion chamber size, integral iron valve guides, and screw-in rocker studs. These heads can be purchased three different ways. Buy a bare pair and use your own valves and hardware, or order the parts and assemble the heads yourself. Of course, you can also order the heads fully assembled and ready to bolt on. Port: 165 cc. Chamber: 72 cc. Valve Sizes: 1.94/1.50. Material: Aluminum.
I Flow @ 0.400: 195 cfm

Valve Lift	Intake	Exhaust w/pipe
0.050	34	22
0.100	60	47
0.200	116	95
0.300	161	126
0.400	195	135
0.500	210	138
0.600	208	139
0.700	210	139
E/I @ 0.400: 69%		

EDELBROCK PERFORMER RPM

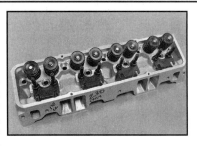

Edelbrock has long enjoyed a reputation for great intake manifolds and now has another hit with this line of aluminum Performer RPM heads. This small 170cc intake port offers outstanding torque potential with good intake and exhaust flow numbers. The exhaust port offers a great E/I percentage of no less than 70 percent and much higher percentages in the low-lift areas, which indicates this head would work well with a single pattern camshaft or a mild dual pattern. Combined with an Edelbrock Performer RPM dual plane and a set of headers, this would make a great combo. We've made 425 hp with a 355ci street small-block using these heads. Port: 170 cc. Chamber: 71 cc. Valve Sizes: 2.02/1.60. Material: Aluminum. I Flow @ 0.400: 218 cfm.

Valve Lift	Intake	Exhaust w/pipe
0.050	31	23
0.100	64	51
0.200	129	104
0.300	181	132
0.400	218	157
0.500	237	173
0.600	227	180
0.700	2301	86
E/I @ 0.400	72%	

DART IRON EAGLE 180

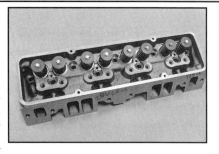

This next-larger, 180cc Dart Iron Eagle head reveals a respectable intake flow curve and an outstanding exhaust port. Again, our flow testing indicates this head would work best with a single pattern camshaft in a 350ci or smaller engine. This 180 Dart head (like all Dart castings) can be purchased either bare or assembled. The 180 head is only offered with 2.02/1.60-inch stainless steel valves, but you can order either straight or angled plugs. Dart also offers this head in either a 64- or 72cc chamber size. All the Dart iron and aluminum heads are drilled with accessory bolt holes in the ends of the heads. Port: 174 cc. Chamber: 68 cc. Valve Sizes: 2.02/1.60. Material: Iron. I Flow @ 0.400: 197 cfm.

Valve Lift	Intake	Exhaust w/pipe
0.050	33	23
0.100	62	52
0.200	121	113
0.300	175	144
0.400	210	167
0.500	209	175
0.600	209	177
0.700	209	—
E/I @ 0.400		79%

Edelbrock E-TEC 170

Given the immense popularity of the iron Chevy Vortec head, Edelbrock has come up with an aluminum high-performance version in two different intake-port volumes. The original Chevy Vortec is patterned after the LT1 intake port, which measured 170 cc when we first tested it ia few years ago. The Edelbrock heads also feature both valve-cover hold-down designs, so they can be used on early engines using standard perimeter-bolt valve covers. The heads also feature polished stainless steel valves and a high-turbulence combustion chamber. The 170 E-TEC head (PN 6097 complete) comes with a 1.94-inch intake and a slightly larger than usual 1.55-inch exhaust valve. These heads can be purchased fully machined and complete or bare and ready to assemble. Keep in mind that these heads require the specific Vortec intake-manifold bolt pattern and intake manifold. Port: 170 cc. Chamber: 64 cc. Valve Sizes: 1.94/1.55. Material: Aluminum. I Flow @ 0.400: 215 cfm.

Valve Lift	Intake Open	Exhaust Open	Exhaust w/Pipe
0.050	32	21	22
0.100	63	48	48
0.200	126	100	104
0.300	177	132	138
0.400	215	160	169
0.500	240	172	180
0.600	240	180	190
E/I @ 0.400		74.4%	78.6%

Flow to Go

Holley 300-570

Holley's latest venture into the small block cylinder head world is the aluminum 175 cc intake-port head. This head is one of three 175 cc intake ports that Holley claims all flow roughly the same. These angle-plug heads employ a 69cc chamber that pumps the compression up when replacing the typical 76cc iron smog heads. These come complete with swirl-polished stainless steel valves and big valvesprings that can handle up to 0.600-inch lift. The combustion chamber is also typical of late-model designs with a kidney shape that increases combustion turbulence with the additional blessing of being less detonation-sensitive. Port: 172 cc. Chamber: 69 cc. Valve Sizes: 2.02/1.60. Material: Aluminum. I Flow @ 0.400: 223 cfm.

Valve Lift	Intake	Exhaust Open	Exhaust w/Pipe
0.050	38	23	23
0.100	72	52	53
0.200	134	96	102
0.300	188	133	143
0.400	223	155	170
0.500	232	158	177
0.600	236	158	180
E/I @ 0.400		69.5%	76%

CATEGORY 2 SMALL-BLOCK HEADS: 180 CC TO 200 CC

BRODIX -8 PRO ALUMINUM

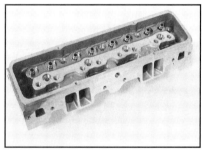

Brodix is one of the leaders in hot performance cylinder heads, but the -8 Pro is perhaps its best-kept street secret. With large 2.08 intake and 1.60-inch exhaust valves, this smaller 181cc intake port flows like gangbusters, with an outstanding 220 cfm at 0.400 inch lift. It quickly tops out at 254 cfm at 0.500 inch lift, but with this much flow, combined with a midsize port, this is an outstanding choice for a strong 355- or 383ci small-block. Brodix offers several versions of this head, either bare or complete with different levels of preparation. If you're looking for an aluminum street head with outstanding potential to crank both horsepower and torque, this Is where to start. Port: 181 cc. Chamber: 67 cc. Valve Sizes 2.08/1.60. I FLow @ 0.400: 220 cfm.

Valve Lift	Intake	Exhaust Open	Exhaust w/pipe
0.050	34	25	25
0.100	63	51	52
0.200	125	91	94
0.300	177	118	123
0.400	220	138	144
0.500	254	155	164
0.600	226	161	170
E/I		63%	73%

HOLLEY SYSTEMAX ALUMINUM

Here's another great cylinder head, this time from Holley. This head also offers outstanding potential, with very similar intake flow numbers to the Brodix -8 head. The Holley looks a little bit better on the exhaust side, but a couple of cfm may not be that critical. Overall, this is another excellent head with great torque potential based on its 186cc intake-port volume. Holley offers this head by itself or in combination with a Holley intake and cam package that makes creating a strong small-block a snap. Port: 186 cc. Chamber: 68 cc. Valve Sizes: 2.02/1.60. I Flow @ 0.400: 222 cfm.

Valve Lift	Intake	Exhaust Open	Exhaust w/pipe
0.050	38	28	28
0.100	66	51	53
0.200	130	90	92
0.300	184	117	122
0.400	222	143	148
0.500	242	168	172
0.600	252	181	187
E/I		64%	73%

Small-Block Chevy Engine Buildups

AIRFLOW RESEARCH
190 ALUMINUM

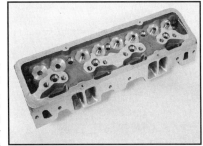

If you're looking for great flow from a great midsize cylinder head, look closely at the AFR 190. This head offers the best flow numbers of all the Category 2 midsize heads with an outstanding exhaust port as well. The AFR 190 head offers great flow throughout the entire valve-lift range, not just at the peak-flow numbers. With a 72-percent E/I relationship, the AFR 190 offers great compatibility with almost any street or mild race combination. There's also a 195cc head that flows virtually the same but employs a larger intake-port opening for Victor Jr.–style intake manifolds. We built a 525 hp 420ci small-block with these heads with mild porting, so the potential is obvious. Port: 191 cc. Chamber: 76 cc. Valve Sizes: 2.02/1.60. I Flow @ 0.400: 244 cfm

Valve Lift	Intake	Exhaust Open	Exhaust w/pipe
0.050	40	31	31
0.100	71	65	67
0.200	144	115	121
0.300	208	146	157
0.400	244	176	188
0.500	262	190	202
0.600	261	197	211
E/I		72%	77%

LGM LT4 ALUMINUM

The LT4 is substantially different from the LT1 casting, and the flow numbers explain why. The head uses larger valves, but the real reason for 230 cfm at 0.400-inch lift can be attributed to the larger intake ports. For a production engine, these are great flow numbers. Like the LT1 engine, these LT4 heads can be used only on a reverse-cooling LT1 engine but do offer some significant horsepower advantages. Port: 195 cc. Chamber: 56 cc. Valve Sizes: 2.00/1.55. I Flow @ 0.400: 230 cfm

Valve Lift	Intake	Exhaust Open	Exhaust w/pipe
0.050	43	37	36
0.100	78	67	67
0.200	147	109	114
0.300	197	133	141
0.400	230	159	167
0.500	250	172	180
0.600	243	177	185
E/I		69%	72%

TRICK FLOW SPECIALTIES
PN 30400001

These new TFS heads were flow-tested and bolted onto a mild 383ci Mouse hooked up to Westech's Superflow 901 dyno. On the dyno we found that flow figures don't tell the whole story because even though these heads didn't flow huge numbers like some of their bigger small-block cousins, they were able to pump up peak engine power by a factor of 15.5 percent just by bolting them on. These may be the best power-for-the-buck small-block aluminum heads currently on the market. For $800 complete, you can't go wrong. Intake port: 195 cc. Chamber: 65 cc. Valve sizes: 2.02/1.60. I Flow @ 0.400: 226 cfm

Valve lift	Intake cfm	Exhaust cfm w/ pipe
0.050	032	028
0.100	065	070
0.200	133	102
0.300	188	140
0.400	226	164
0.500	250	183
0.600	N/A	N/A
I/E @ 0.400 lift		73%

Air Flow Research 180

This is AFR's smaller version of the company's killer 190- to 195cc street heads. This head is intended for outstand-ing mid-range torque in 350ci engines, and would also work extremely well with a smaller-displacement engine like a 302ci, 305ci, or 327ci. The flow numbers are especially impressive on the exhaust side of this head, leading us to assume that these heads would perform better with a single-pattern camshaft as opposed to a dual-pattern cam. The exhaust-to-intake relationship (E/I) is an outstanding 84 percent using exhaust-flow numbers with a pipe attached or 76 percent without the pipe. There are only a handful of heads that outflow the AFR 180 at 0.400-inch intake lift, and all are larger than the AFR 180. This makes this head extremely impressive for its size and worthy of consideration. One key to the strong flow numbers is that it is a fully CNC-machined intake port. Port: 181 cc. Chamber: 68 cc. Valve sizes: 2.02/1.60. Material: Aluminum. I Flow @ 0.400: 230 cfm.

Valve Lift	Intake Open	Exhaust Open	Exhaust w/Pipe
0.050	35	24	25
0.100	66	55	55
0.200	137	107	115
0.300	193	152	164
0.400	230	175	193
0.500	250	186	208
0.600	255	190	214
E/I @ 0.400		76%	84%

Edelbrock E-TEC 200

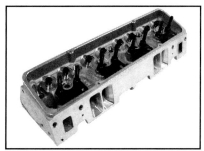

This is the larger version of Edelbrock E-TEC's pair of aluminum small block Vortec head upgrades (PN 6098 complete). These heads increase the intake-port volume from 170 to 200 cc and come with 2.02/1.60-inch valves. Like the smaller E-TEC heads, these also require the Vortec-style intake manifold. These heads also feature a 64cc chamber, placing the compression around 10:1 with a flat-top piston in a 350ci small-block. This larger intake port does outflow the stock Vortec iron head, but only at valve lifts above 0.500 inch. Like the smaller 170cc E-TEC head, this head does have a dramatically better exhaust port that will help performance throughout the entire power curve. Port: 200 cc. Chamber: 64 cc. Valve Sizes: 2.02/1.60. Material: Aluminum. I Flow @ 0.400: 220 cfm.

Valve Lift	Intake Open	Exhaust Open	Exhaust w/Pipe
0.050	30	23	23
0.100	63	50	50
0.200	123	102	106
0.300	177	138	144
0.400	220	171	179
0.500	252	188	194
0.600	257	197	206
E/I @ 0.400		78%	81%

Small-Block Chevy Engine Buildups

CATEGORY 3 SMALL-BLOCK HEADS: 200 CC AND UP

TFS TWISTED WEDGE ALUMINUM

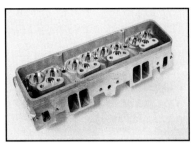

The Twisted Wedge small-block head is a different animal. In an attempt to improve airflow, TFS changed the intake-valve angle from the standard 23 degrees to a shallower 13 degrees. As you can see from the flow at 0.400 inch, this head works very well. Ironically, the Twisted Wedge's exhaust port is what really shined in our testing. While the intake port is certainly good, at 0.400-inch valve lift the E/I relationship is an outstanding 77 percent. Only the AFR 210 head runs equal to the TFS exhaust port. Like many heads such as the Performer and Brodix -8 Pro, the Twisted Wedge comes with a mild amount of hand work to the short-side radius of the seats. Port: 200 cc. Chamber: 64 cc. Valve Sizes: 2.02/1.60. I Flow @ 0.400: 227 cfm.

Valve Lift	Intake	Exhaust Open	Exhaust w/pipe
0.050	35	39	39
0.100	73	67	69
0.200	127	113	116
0.300	186	149	155
0.400	227	176	183
0.500	254	202	204
0.600	265	211	213
E/I		70%	80%

WORLD PRODUCTS SPORTSMAN II IRON

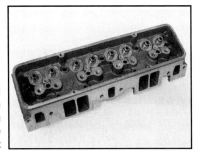

The World Products iron 200cc cylinder head has been around for a while and enjoys a reputation as a reliable cylinder head that makes good power. The 225-cfm flow number at 0.400-inch valve lift is right there in the running, and the price of these heads is especially attractive. There are tons of options for these heads as well, with short-side radius-porting work available that can improve the airflow. While these heads are heavy, even for iron castings, they are durable and especially affordable. The E/I relationship is a little weak at 0.400-inch valve lift, but this is easily recitified with a dual-pattern camshaft. The Sportsman II head also features dual-valve-cover bolt patterns and offers a 50-state legal E.O. that is allowed to be used on late-model emissions engines as well. Port: 201 cc. Chamber: 72 cc. Valve Sizes: 2.02/1.60. I Flow @ 0.400: 225 cfm.

Valve Lift	Intake	Exhaust Open	Exhaust w/pipe
0.050	40	28	28
0.100	68	55	57
0.200	132	98	102
0.300	187	123	131
0.400	225	138	150
0.500	240	150	164
0.600	243	156	162
E/I		61%	67%

GM LS1 ALUMINUM

It was surprising to see the LS1 cylinder head in Category 3, but the burette doesn't lie. The tall cathedral intake ports displace a whopping 204 cc. Even with small 2.00/1.55-inch valves, this head still moves some air, with a 222-cfm port flow at 0.400-inch valve lift. It also has a decent, 70-percent E/I relationship at 0.400-inch valve lift. The other major departure from typical small-block style is the LS1's 15-degree valve angle, which promotes a much flatter and more efficient combustion chamber. While little has been mentioned about combustion chambers, this is a place where good design can create additional power without causing detonation. It's worth comparing the chamber shape of this newest Chevrolet cylinder head with the older 462 and 882 chambers. Note how the new chamber is kidney-shaped to help direct combustion around the exhaust valve. While obvious, it's important to mention that the LS1 head will not interchange with any other small-block cylinder head. Port: 204 cc. Chamber: 65 cc. Valve Sizes: 2.00/1.55. I Flow @ 0.400: 222 cfm.

Valve Lift	Intake	Exhaust Open	Exhaust w/pipe
0.050	46	27	31
0.100	83	55	62
0.200	134	98	104
0.300	189	130	136
0.400	222	156	165
0.500	240	175	179
0.600	252	182	185
E/I		61%	67%

EDELBROCK VICTOR JR. ALUMINUM

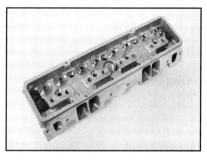

The Edelbrock Victor Jr. head came out looking good with flow of over 250 cfm at 0.500-inch valve lift and a great flowing exhaust port. The E/I relationship stands at 81 percent. The combustion-chamber design is the classic kidney shape that enhances combustion efficiency without the need for massive amounts of timing. We actually tested this head all the way up to 0.700-inch of valve lift where the Victor Jr. cranked out 284 cfm, which shows that this port would work really well with a roller camshaft and lots of lift. Port: 215 cc. Chamber: 68 cc. Valve Sizes: 2.08/1.60. I Flow @ 0.400: 219 cfm.

Valve Lift	Intake	Exhaust Open	Exhaust w/pipe
0.050	35	28	28
0.100	69	57	57
0.200	123	108	111
0.300	173	141	145
0.400	219	178	180
0.500	255	194	198
0.600	275	203	206
0.700	284		
E/I		81%	82%

AIRFLOW RESEARCH 210 ALUMINUM

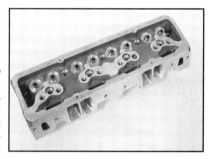

When you get into these larger heads, the flow really starts to take off. The AFR heads measured a little larger than 210 cc, but when you've got 240 cfm at 0.400-inch valve lift, who cares? The AFR 210 rocked to a max flow of 284 cfm at 0.700-inch valve lift right out of the box. With its E/I of 75 percent at 0.400-inch valve lift, it's hard to find something wrong with these heads. As street heads they're probably a bit large for a 350ci small-block, but they'd be a killer for a larger-displacement Mouse motor, such as a 408- or 420ci thumper. When it comes to airflow, especially around 0.400- and 0.500-inch valve lift, the AFR is unparalleled among the heads we tested. Port: 215 cc. Chamber: 67 cc. Valve Sizes: 2.08/1.60. I Flow @ 0.400: 240 cfm

Valve Lift	Intake	Exhaust Open	Exhaust w/pipe
0.050	33	36	36
0.100	67	72	73
0.200	140	115	118
0.300	197	148	153
0.400	240	180	183
0.500	271	198	200
0.600	279	210	210
0.700	284	218	218
E/I		75%	76%

BRODIX TRACK 1 ALUMINUM

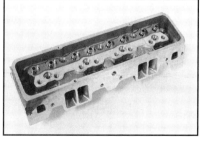

The Brodix Track 1 head was delivered with the seats installed and machined with a minor amount of work below the seat. Experience has proven that a small amount of short-side radius work within half an inch of the valve seat can have an amazing effect on airflow. The Track 1 is also available from Brodix in several valve-size configurations. This head enjoys a solid reputation in the industry, and while the flow at 0.400 is down slightly, the numbers at 0.500 are certainly competitive. Port: 216 cc. Chamber: 67 cc. Valves Sizes: 2.08/1.60. I Flow @ 0.400: 209 cfm.

Valve Lift	Intake	Exhaust Open	Exhaust w/pipe
0.050	25	23	23
0.100	60	47	48
0.200	115	100	106
0.300	166	138	147
0.400	209	171	178
0.500	246	180	191
0.600	267	186	200
0.700	266		
E/I		82%	76%

Small-Block Chevy Engine Buildups

CANFIELD ALUMINUM

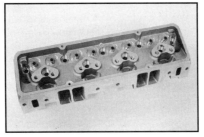

The Canfield is a relative newcomer to the cylinder head market, but with its 220cc intake port, it certainly gets with the program. The head is only available with the valves supplied, and with 247 cfm at 0.400-inch valve lift and 258 cfm at 0.500-inch valve lift, these heads can make power. One area to look at with the Canfield is the excellent low- and midlift flow numbers. The Canfield outflows all the big heads from 0.050- through 0.400-inch valve lift, which makes this a great head for bigger-cubic-inch small-blocks where outstanding torque without loss of horsepower is required. The combustion chamber on the Canfield is also completely CNC-machined. Port: 220 cc. Chamber: 67 cc. Valve Sizes 2.02/1.60. I Flow @ 0.400: 247 cfm.

Valve Lift	Intake	Exhaust Open	Exhaust w/pipe
0.050	38	33	33
0.100	74	61	63
0.200	141	104	107
0.300	201	135	143
0.400	247	164	175
0.500	258	175	190
0.600	257	183	200
0.700			
E/I		66%	71%

DART CONQUEST ALUMINUM

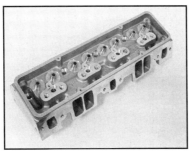

The Dart Conquest is the latest addition to the large volume aluminum performance cylinder head market. While the flow numbers appear to be slightly low at 0.400-inch valve lift, the flow at 0.500-inch valve lift makes a huge jump to 252 cfm, and by 0.600 inch the flow is pumping 274 cfm. With an E/I relationship at 0.400-inch valve lift of 72 percent and 68 percent at 0.500-inch valve lift, this head is certainly in the running. This head comes machined for both center-bolt and perimeter-style valve covers. The Conquest can be optioned with either the large 220cc port or a smaller 200cc port size. Valve sizes can also be ordered up to 2.08/1.625 inches and in either 64cc or 72cc chamber sizes. Port: 220 cc. Chamber: 62 cc. Valve Sizes: 2.05/1.60. I Flow @ 0.400: 217 cfm.

Valve Lift	Intake	Exhaust Open	Exhaust w/pipe
0.050	38	26	27
0.100	73	58q	55
0.200	126	108	105
0.300	171	134	133
0.400	217	158	162
0.500	252	168	179
0.600	274	173	188
0.700	268	178	195
E/I		72%	74%

DART IRON EAGLE

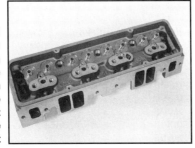

The other new head from Dart is the Iron Eagle, which also has a 220cc intake port. The Iron Eagle offers a budget alternative to the aluminum Conquest cylinder head. While similar to the Conquest head, the flow numbers are relatively strong through the midrange airflow with 258 cfm at 0.500-inch valve lift. It offers a 70 percent E/I relationship at 0.400-inch of valve lift. This head also comes machined for both center-bolt and perimeter-style valve covers. Like the Conquest aluminum, port-size options include the 200- or 220cc ports, and 2.02- or 2.05-inch intakes. Port: 220 cc. Chamber: 72 cc. Valve Sizes: 2.05/1.60. I Flow @ 0.400: 226 cfm.

Valve Lift	Intake	Exhaust Open	Exhaust w/pipe
0.050	28	29	29
0.100	65	58	59
0.200	129	108	112
0.300	182	134	139
0.400	226	158	166
0.500	258	168	178
0.600	255	173	186
0.700		175	187
E/I		70%	73%

CANFIELD PN 23600 ALUMINUM

Canfield offers this head with two different intake runner volumes, but we were only able to test the big 220cc head at this time. This race-oriented small-block head requires the use of a 0.150-inch offset intake rocker arm to accommodate the repositioned intake valve. The practice of spreading the valves farther apart to increase valve diameter and airflow is becoming more common as intake ports get larger in the search for higher flow figures. The shape of the Canfield's CNC'd combustion chambers is unusual but will still accommodate most production piston domes on high-compression engines. These heads came equipped with a big 2.08-inch intake valve and a 1.60-inch exhaust valve with the area directly above the valve faces blended smooth in both ports. We think this head would work great on a 383ci-or-larger small-block intended for heavy street/strip action. Port: 223 cc. Chamber: 65 cc. Valve sizes: 2.08/1.60. I Flow @ 0.400: 245 cfm.

Valve lift	Intake cfm	Exhaust cfm w/ pipe
0.050	034	028
0.100	069	061
0.200	146	114
0.300	206	144
0.400	245	174
0.500	260	182
0.600	266	190
I/E @ 0.400 lift	71%	

BRODIX -1X

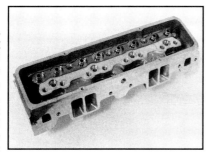

The Brodix Track 1 cylinder head has long been the standard against which many small-block heads have been measured. Brodix has taken that head to the next level with its -1X. The new 1X features 40/60 valve spacing, meaning the centerline of each valve has been moved away from its standard location (0.040 inch on the exhaust and 0.060 inch on the intake) to accommodate a larger-diameter valve and move both valves away from the chamber walls for unshrouding. This means that you'll need special, offset shaft-mounted rocker arms, which are available from Brodix or T&D. All other standard small-block components will fit these 23-degree heads. If purchased as a complete package, the -1X heads receive a bowl blend, a CNC intake port match, and CNC-machined chambers. Port: 223 cc. Chamber: 67 cc. Valve sizes: 2.10/1.60. I Flow @ 0.400: 251 cfm

Valve lift	Intake cfm	Exhaust cfm w/ pipe
0.050	027	023
0.100	066	051
0.200	140	100
0.300	202	131
0.400	251	151
0.500	278	171
0.600	271	180
I/E @ 0.400 lift		60%HP

DART IRON EAGLE 200 cc

The midsize Dart Iron Eagles could become one of the more popular performance iron heads since it can work well with virtually any 350ci or larger small-block. This head offers decent midrange flow numbers on the intake side with an exhaust port offering great E/I flow from 0.050-inch up to 0.300-inch lift. Dart offers the 200 in the widest range of options including 2.02-, 2.05-, and 2.08-inch intake valve sizes with exhaust choices including 1.60 and 1.625 inches. Straight or angle plugs are available, as are either 64 or 72cc chambers. While we tested the Dart Iron Eagles, the Pro 1 series of Dart aluminum heads offers duplicate port configurations. This means the flow numbers should be the same between the Iron Eagle and Pro 1 heads offered in the 200 and 215cc sizes. Port:200 cc. Chamber: 65 cc. Valve Sizes: 2.02/1.60. Material: Iron. I Flow @ 0.400: 212 cfm.

Valve Lift	Intake	Exhaust w/pipe
0.050	31	25
0.100	61	53
0.200	122	105
0.300	174	126
0.400	212	164
0.500	234	174
0.600	236	175
0.700	239	177
E/I @ 0.400		77%

Small-Block Chevy Engine Buildups

DART Pro 1 215 cc

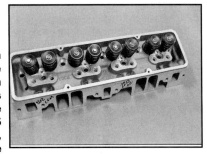

Since the aluminum Pro 1 heads are the same as the iron pieces as far as ports are concerned we tested the Pro 1 215 cc head. Interest-ingly, the particular port we measured actually measured over 220 cc, which is probably due to production tolerances and not a typical port volume for these heads. The 215 offered a reasonable intake flow curve with over 250 cfm of airflow using a 2.05-inch intake valve. The exhaust-to-intake relationship averages well into the mid-70 percentile, which is borderline whether you would use a single or dual pattern camshaft. This is perhaps the most versatile head with tons of options in valve sizes, chamber size, and straight or angle plug. This head would best be applied to a high-winding 377, 383, or 406ci small-block. Port: 221 cc. Chamber: 65 cc. Valve Sizes: 2.05/1.60. Material: Aluminum. I Flow @ 0.400: 224 cfm.

Valve Lift	Intake	Exhaust w/pipe
0.050	34	28
0.100	66	55
0.200	128	100
0.300	182	136
0.400	224	162
0.500	253	173
0.600	256	176
0.700	265	178
E/I @ 0.400		72%

BRODIX -11X 220 cc

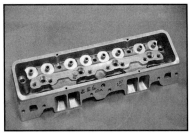

Brodix offers this large-port, 23-degree valve angle small-block head as one of its largest ports in the stock valve-angle lineup. Intake port flow at 0.400-inch lift is a reasonable 220 cfm, which is right there with comparably sized intake ports. The exhaust port flows much better at the higher lift numbers than at low lifts if you do the exhaust-to-intake comparisons. Low-lift flow hits a low of 61 percent at 0.200-inch lift but then charges back with 79 percent at max lift. This is just one of dozens of Brodix small-block heads. Even within the -11X line there are several variations offered, including a minor or complete port work that can increase intake port flow up to 300 cfm at 0.600-inch lift. The -11X is also available in a raised port configuration. Port: 220 cc. Chamber: 68 cc. Valve Sizes: 2.08/1.60. Material: Aluminum. I Flow @ 0.400: 220 cfm.

Valve Lift	Intake	Exhaust w/pipe
0.050	34	25
0.100	68	45
0.200	127	78
0.300	178	116
0.400	220	157
0.500	249	182
0.600	249	197
0.700	256	204
E/I @ 0.400		71%

DART IRON EAGLE 230 cc

This is the largest of the Dart Iron Eagle series of 23-degree small-block heads. This head is best used with a large displacement small-block at 400 ci or larger along with a camshaft that can offer lifts above 0.500-inch to take advantage of the head's great high-lift flow numbers. The exhaust-to-intake relationship is acceptable with 76 percent at 0.400-inch valve lift. As with all the Dart Iron Eagles, the castings are excellent quality and fitted with hardened exhaust seats. Most of the heads we tested had an excellent transition between the valve seats and the port throat that did not appear to require serious hand work to improve low-lift flow. Port: 229 cc. Chamber: 65 cc. Valve Sizes: 2.08/1.600. I Flow @ 0.400: 218 cfm.

Valve Lift	Intake	Exhaust w/pipe
0.050	32	25
0.100	61	53
0.200	119	102
0.300	171	144
0.400	218	167
0.500	254	177
0.600	272	178
0.700	275	179
E/I @ 0.400		76%

Pro Action Iron Lightning 220

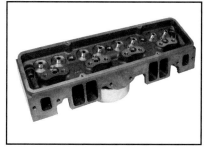

The trend in small-block heads is definitely toward the larger intake ports. This Pro Action Iron Lightning head is one of four different heads available through Air Flow Research (AFR). By the time you read this, AFR will have 180-, 200-, 220-, and 235cc intake-port heads available. The only heads available at press time were the two larger versions. These heads are available with either 64 or 72cc chambers and either as angle or straight plug designs to clear your headers.

Port: 225 cc. Chamber: 67 cc. Valve sizes: 2.05/1.60. Material: Iron. I Flow @ 0.400: 207 cfm.

Valve Lift	Intake	Exhaust Open	Exhaust w/Pipe
0.050	28	24	24
0.100	60	51	52
0.200	115	98	102
0.300	161	128	132
0.400	207	155	160
0.500	239	171	176
0.600	257	181	187
E/I @ 0.400		75%	77%

Pro Action Iron Lightning 235

These are the largest of the Pro Action Iron Lightning lineup. The heads are claimed to be 235 cc, but we measured several ports and came up with 245 cc, which makes them the largest 23-degree-valve-angle intake port we've tested. Like its smaller cousin, this head is available from AFR with 64cc or 72cc chambers and in either straight or angle plug design. With ports this huge, this head would have to be used on a 406ci-or-larger small-block or a very-high-rpm smaller engine. Port: 245 cc. Chamber: 64 cc. Valve sizes: 2.08/1.60. Material: Iron. I Flow @ 0.400: 204 cfm.

Valve Lift	Intake	Exhaust Open	Exhaust w/Pipe
0.050	32	26	25
0.100	55	53	53
0.200	116	100	103
0.300	159	127	131
0.400	204	156	158
0.500	242	172	176
0.600	270	183	186
0.700	291	188	190
E/I @ 0.400		76.4%	77.4%

World Products Motown 220 Iron

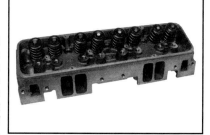

World Products' first big head was this iron 220cc small-block head. The numbers are impressive in the mid-lift areas, placing it in the lead pack of Category 3 small-block heads. The heads we tested were supplied with 2.08-inch valves while the catalog claims 2.05-inch valves, so there is probably a flow gain here as well. Port: 217 cc. Chamber: 65 cc. Valve sizes: 2.08/1.60 Inches. I flow @ 0.400: 222 cfm.

Valve Lift	Intake	Exhaust Open	Exhaust w/Pipe
0.050	29	23	23
0.100	60	49	50
0.200	129	100	106
0.300	183	133	142
0.400	222	158	165
0.500	244	169	176
0.600	254	174	184
0.700	259	177	188
E/I @ 0.400		71%	74%

Small-Block Chevy Engine Buildups

World Products Motown 220 Aluminum

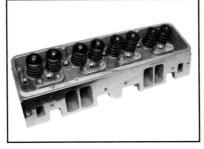

The aluminum head flows much better than the iron 220. In fact, this head is right in the heavy-hitter league with the AFR 210 and the Canfield 220 heads. For example, the Motown 220 has good mid-lift flow numbers in the 0.300- and 0.400-lift range with a reasonable exhaust port. At 0.500-inch lift, this head outflows both the AFR and Canfield, though only by a slim margin. Port: 224 cc.

Chamber: 64 cc. Valve sizes: 2.08/1.60 inches. I flow @ 0.400: 244 cfm.

Valve Lift	Intake	Exhaust Open	Exhaust w/Pipe
0.050	32	22	22
0.100	58	52	52
0.200	132	101	105
0.300	195	153	156
0.400	244	174	181
0.500	274	183	192
0.600	277	185	194
0.700	272	185	195
E/I @ 0.400		71%	74%

Trick Flow Specialties Aluminum 215

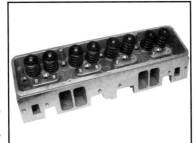

The latest 23-degree aluminum head from TFS sports a 215cc intake port and offers good flow potential. Delivering 187-cfm flow at 0.300-inch lift, this head exhibits great mid-range flow that is only exceeded by the AFR, Brodix, Canfield, and Dart Pro 1 CNC heads, which all sport larger ports. While a 215cc head might be a bit big for a 350ci engine, it's primed for a thumpin' 383ci or 406ci small-block.

Port: 219 cc. Chamber: 66 cc. Valve sizes: 2.08/1.60 inches. I flow @ 0.400: 227 cfm.

Valve Lift	Intake	Exhaust Open	Exhaust w/Pipe
0.050	33	24	24
0.100	67	56	56
0.200	132	103	107
0.300	187	141	148
0.400	227	171	177
0.500	247	187	194
0.600	257	195	204
0.700	263	202	210
E/I @ 0.400		75%	78%

Dart Pro 1 CNC Aluminum

This is Dart's entry into the CNC-ported world of big street heads, combining excellent mid-lift flow at 0.300- and 0.400-inch lift, and an outstanding exhaust port that delivers 80 percent E/I. This is the head we used on our 406ci small-block in Chapter 14 that made 515 hp. Port: 220 cc. Chamber: 64 cc. Valve sizes: 2.08/1.60 inches. I Flow @ 0.400: 236 cfm.

Valve Lift	Intake	Exhaust Open	Exhaust w/Pipe
0.050	36	23	26
0.100	71	62	68
0.200	144	103	112
0.300	193	139	147
0.400	236	185	189
0.500	270	211	215
0.600	292	220	224
0.700	293	224	229
E/I @ 0.400		78%	80%

Flow Power

Choosing the Right Cylinder Heads
By Jeff Smith
Photography by Jeff Smith

18

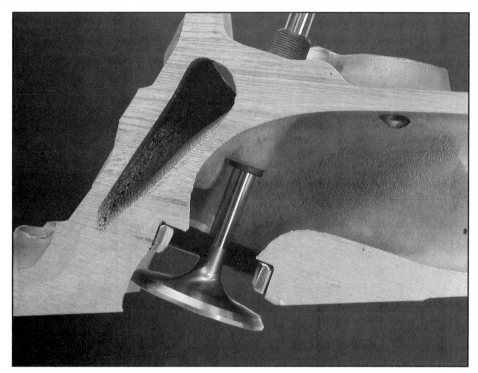

It's all about choices. There are probably over 100 different small-block cylinder heads on the market today. Narrow that down to heads for street use and that still leaves a river of heads to wade through when it comes time to pick the right heads for your next street small-block.

The good news about aftermarket small-block heads is that almost anything you buy will be better than those old stock iron castings. But there are good choices and then there are better choices. Typically, the decision will more likely be determined by the thickness of your wallet than by any flow evaluation. *Chevy High Performance* has been assembling flow data on cylinder heads since 1998. In that time, we've flow tested over 35 different small-block cylinder heads on Westech Performance's SuperFlow flow bench. In fact, all the heads that CHP has tested can be found on our Web site, www.chevyhiperformance.com, and of course, a lot of it is in the last chapter too. With all this information, we've had a chance to study these flow numbers at length. While the flowbench is not the ultimate arbiter of a great cylinder head, the bench is the best indicator short of outright dyno-testing.

The biggest problem with flow bench data is that everyone immediately uses the cfm numbers at the highest valve-lift point to compare cylinder heads. However, it's pointless to use a 0.600-inch valve-lift flow number to decide on a cylinder head when the maximum valve lift you will ever see on your street motor is 0.500 inch.

The smarter move is to evaluate the mid-lift flow numbers between 0.200- and 0.400-inch valve lift. These numbers represent the bulk of a cylinder head's true flow potential. This includes exhaust flow numbers as well as intake.

Another important factor is the intake port volume. This is an area where there is plenty of room for choice. Standard 23-degree small-block (excluding the LT1, LT4, and LS1) cylinder heads vary in port volume from 160 to 170cc stock to as large as 230cc. This is a huge port-volume range. Generally, the larger heads are designed for use on high-horsepower, high-rpm engines or larger-displacement 406ci through 430ci small-blocks.

The rule of thumb is to choose a cylinder head that offers the best flow with the smallest port volume. For example, if you can find a 190cc intake-port-volume head that flows the same as a 220cc head, the smaller port head would be a better choice since the smaller port contributes to higher port velocity, which improves torque.

Of course, you must also look at the exhaust port in relationship to the

Small-Block Chevy Engine Buildups

Combustion-chamber design can have a significant effect on performance. The kidney shape, such as on this TFS chamber, creates additional turbulence that improves efficiency and allows more compression without detonation.

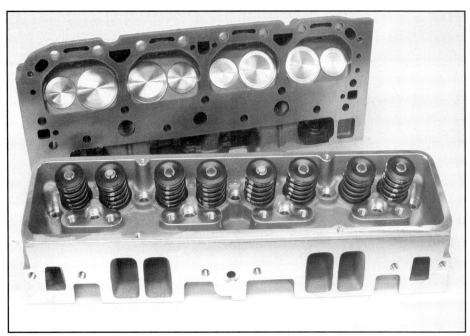

Aluminum heads offer several advantages over iron including light weight and ease of repair. Most aluminum heads are more expensive than iron, but the power is usually worth the expense.

Pocket porting a stock replacement head like the World S/R Torquer is an excellent way to improve flow at a low cost.

intake. The best way to evaluate this is to divide the exhaust-port cfm by the intake-port cfm at the same valve lift.

For example, an exhaust-port flow of 152 cfm at 0.300-inch valve lift divided by an intake port with 193-cfm flow at the same lift will give you an exhaust-to-intake relationship expressed as a percentage (E/I). In this example, 152/193 = 79 percent E/I, which is a good number. E/I numbers between 75 and 85 percent are considered very good numbers.

But you have to be careful because an under-performing intake port can make an exhaust port look good.

Price, flow potential, intake port volume, E/I percentages, iron versus aluminum, compression ratio, and several other variables contribute to the makings of a great cylinder head. The good news is that there are tons of cylinder heads from all the major cylinder-head companies that offer a tremendous selection. Bigger is not always better when it comes to a strong street small-block cylinder head. The best news is that all the cylinder heads in this story represent a solid performance investment for a street small-block. All you have to do is decide which is best for your application.

Go with the Mid-Lift Flow

When comparing cylinder-head flow data, it seems that everyone always looks at the maximum-lift flow data. But especially for street engines, the smarter move is to look at the mid-lift flow numbers. We decided to look more closely at the more popular iron and aluminum

Flow Power

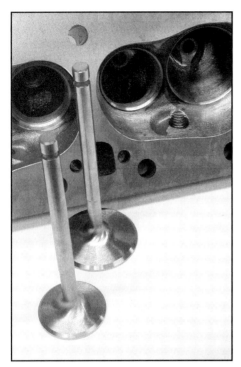

A valve job can have a major effect on flow potential. It's possible to improve flow with simple modifications to the seat angles. Larger and better aftermarket valves can also improve flow on a stock or replacement head.

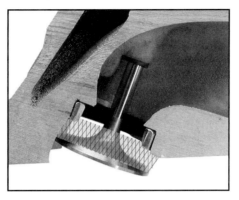

The area created by the circumference of the valve diameter and the valve-lift height creates a valve "window" that helps determine airflow. All kinds of variables affect this flow, which is why some heads flow better than others.

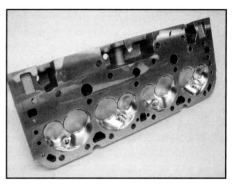

An excellent aluminum street head is the TFS 23-degree small-block head. This head combines outstanding mid-lift flow numbers with an affordable price just under $1,000. This head comes with a 64cc chamber, but an engine with aluminum heads can usually handle a 10:1 compression ratio without detonating on 92-octane pump gas.

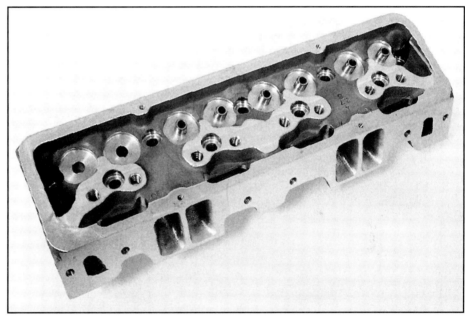

If you've looked at all the flow numbers for 180–200cc intake-port heads, it's easy to pick out the exceptional flow characteristics of the 190/195cc Air Flow Research head. While expensive, the AFR 190/195 is the best flowing head for its size that we've tested.

small-block heads at the 0.300-inch valve lift.

Why 0.300-inch valve lift? According to knowledgeable engine gurus like the late Smokey Yunick and Zora Arkus-Duntov, concentrating on valve lifts at 65 to 68 percent of maximum valve lift contributes to both a strong overall torque curve and good horsepower. Number crunching 65 percent of 0.480-inch valve lift equals 0.312 inch.

We used this information to evaluate several small-block cylinder heads that we've flow-tested over the last four years. Since many of the heads that we tested used varying intake-valve sizes (from 1.94 to 2.08 inches), we multiplied the circumference of the valve [diameter times 3.1417 (pi)] times the valve lift to create a square inch area "flow window." Then we divided this valve window area by the flow at 0.300-inch valve lift. By doing this, we eliminated the variable of valve diameter by creating a flow-per-square-inch criterion.

Here's how this works using the GM Performance Parts Vortec iron cylinder head. The Vortec head uses a 1.94-inch intake valve to flow 190 cfm at 0.300-inch lift. We multiplied 1.94 inches times pi = 6.094 inches of circumference. This figure times the valve lift equals a flow window area (6.094 x 0.300 = 1.828 square inches). Then we divided the flow by the area (190/1.828 = 104 cfm per square inch). This number by itself really doesn't mean much. But when we compared over 15 cylinder heads using this evaluation, the Vortec ranked fourth overall. This is just further reinforcement for how well this cylinder head flows between 0.100 and 0.400 inch of valve lift.

Small-Block Chevy Engine Buildups

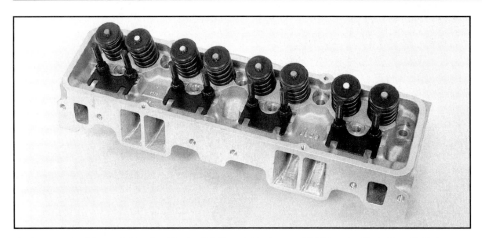

We've also had powerful luck with the Edelbrock Performer RPM head. We built a 355ci motor that made an excellent 425 hp with the Performer RPM heads, Performer RPM intake, and a mild hydraulic cam with 219 degrees of intake duration.

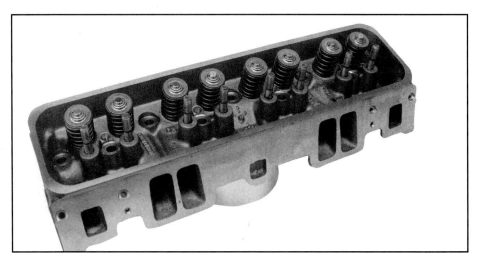

In our opinion, the Vortec head from GM Performance Parts offers the best power value for your dollar. Turn to the next chapter to see how to bolt one on.

The following list includes the top 15 cylinder heads that flowed the best when averaged between 0.300 and 0.400 inch of valve lift. We should also state that all of these heads perform well. Should this mid-lift flow data be the only criterion for choosing a cylinder head for a 350ci/450 hp street engine? The simple answer is no because it's dangerous to base a decision on only one data point. But we would certainly encourage the use of this information as part of an overall decision on your next set of cylinder heads. Our lawyer has also advised us to say that your mileage may vary and the secretary will disavow any knowledge of our actions should we be caught. So there.

The 300 Club
1. AFR 190
2. Canfield 220
3. Canfield 195
4. Vortec iron
5. AFR 180
6. AFR 210
7. TFS 195
8. Holley 300-570
9. World Sportsman II 200
10. Dart Iron Eagle 220
11. Edelbrock Performer RPM
12. World S/R Torquer
13. Brodix -8
14. Stock 882 iron
15. Edelbrock Victor Jr.

Best Buy

If we had to choose one cylinder head out of all the heads listed in this chapter as a CHP best buy it would have to be the GM Performance Parts iron Vortec head. Simply put, the Vortec is at or near the top of every selection criterion we've included in this chapter.

The Vortec offers outstanding mid-lift airflow numbers but it also has some limitations. The 64cc combustion chamber can create a rather steep 10:1 compression ratio when used with a flat-top piston in a 350ci or larger small-block.

The Vortec requires a specific Vortec intake manifold, which means you will probably have to purchase a new intake. The valvetrain is limited to around 0.470-inch valve lift unless you modify the valve guides and buy new springs to allow added retainer-to-seal clearance for a more aggressive camshaft. Finally, the Vortec requires using a late-model center-bolt style valve cover rather than the more typical perimeter bolt arrangement. Even with all these limitations, the Vortec head still gets our vote for the best buy in the mild street small-block cylinder head category.

The iron Vortec offers the best bang for your bucks when it comes to a complete, ready-to-bolt-on iron small-block Chevy cylinder head. This head is capable of 390 to 420 hp when combined with a performance cashaft. Turn to the next chapter for more information on this head.

Bolting On Vortec Heads
How To Install GM's Best Budget Performance Heads
By Jeff Smith
Photography by Jeff Smith

19

Horsepower is one thing, but what really defines performance is when you make serious power on a brown-bag budget. Nothing impresses the troops more than 400 hp for much less than the next guy. One of the star players in this budget approach is the GM Performance Parts' small-block Vortec cylinder head, which we recommended so highly in our last chapter, as well as in many previous articles of Chevy High Perfomance. But there seems to be a bit of confusion about the best way to bolt these heads on an engine. Let's take a look at the details.

The Vortec head's power potential is almost legendary. This head was used on late-model Chevy L31 truck engines and created by stuffing the LT1 Corvette ports in an iron casting. With excellent airflow capability bone stock, the Vortec is not only very competitively priced at usually just under $450 per pair complete with springs and valves, but also makes excellent power. But there are three areas that need to be addressed before you bolt these babies on.

The first hurdle that you must clear is the intake manifold question. The Vortec head uses the less-popular eight-bolt intake manifold pattern as opposed to the more standard 12-bolt pattern. The Vortec intake ports are also quite a bit taller than a standard small-block Chevy port. GM Performance Parts and Edelbrock offer several different intakes in both single- and dual-plane configurations. Unfortunately, these manifolds are more expensive than standard small-block intakes, which reduces some of the cost benefit of the heads.

Some circle-track racers have also successfully experimented with drilling and tapping the Vortec head to use the more common small-block Chevy intake manifolds. While this does work, the Vortec's intake flange area is rather thin in this area, which may cause sealing problems when used in high-mileage applications. Another important consideration is the height of the Vortec's intake port. A dual-plane intake like the Edelbrock Performer will not have enough material at the top of the port to properly seal the port. The intake manifolds designed for the Vortec head have this additional sealing material and will not present a problem. While you can drill these Vortec heads for the old-style intake manifold, the best approach is to use the dedicated intake manifolds, even if it does require spending more money.

The second area to address is compression ratio. We've included various compression-ratio scenarios, but with a flat-top piston 355ci using a typical 0.020-inch deck height and a 0.038-inch-thick head gasket (like the 1003 Fel-Pro), the compression is a stout 9.75:1. Even with a 0.051-inch-

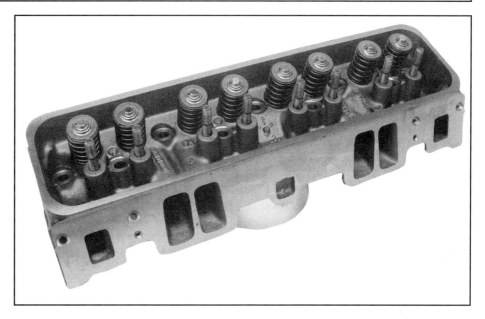

When you buy a Vortec iron head complete, this is what you get. The heads come with press-in studs and rather weak valvesprings. If you don't do anything else, converting to screw-in studs and better springs makes this a more durable head.

Small-Block Chevy Engine Buildups

The Vortec head comes with a 64cc chamber with 1.94/1.50-inch valves and an induction-hardened exhaust valve seat.

The stock 1.250-inch od valvesprings need to be changed, but be careful. Many high-perf springs may not clear the rather large 0.850-inch od valve-guide boss. This can be machined down.

GM Performance Parts offers this slick adapter that allows the use of the standard small-block perimeter-bolt valve cover. Unfortunately, this billet-aluminum adapter is expensive. The alternative is centerbolt valve covers that are available through several aftermarket companies like Proform, TD Performance, and others.

thick head gasket (like the one from ROL), the compression is 9.5:1, which is on the high side of what's preferable in order to avoid detonation difficulties. A longer duration camshaft will help in a case like this, but would not be a good idea if this engine were intended as a daily driver. In this case, the best solution is a dished piston that would bring the compression down to around 9.0:1. Larger engines such as a 383 or 400 will require a dished piston to keep the compression at around 9:1.

One thing to keep in mind with the Vortec head is that the combustion chamber is much more efficient than the older 64cc bathtub style chambers. This new chamber requires less ignition timing, which means it's possible to run a slightly higher static-compression ratio with this head and still not need more than perhaps 34 to 36 degrees of total ignition lead. With a short-duration cam, this may require less initial timing and more mechanical advance, but that's an easy fix. For example, our buddy Tim Moore is building a 355 with these heads and a pair of dished hypereutectic Federal-Mogul pistons. With a zero deck, a 0.041-inch gasket, and a 12cc piston dish, the compression comes out to 9.55:1. He's using Chevy's HOT cam (218/228 @ 0.050 with 0.525-inch valve lift and a lobe separation angle of 112 degrees) that should make for an excellent street engine.

The third situation is somewhat minor, but the Vortec valvesprings are nothing more than weak sister stockers that are dangerously close to coil bind with valve lifts of more than 0.475 inch. Chevy offers stronger valvesprings that are better suited to more rpm and more lift and will clear the inside valve-guide boss. The Chevy spring is 0.070 inch larger in outside diameter (1.320 versus 1.250 inch), which will require enlarging the spring-seat width on the head. This is best done before bolting the heads on the engine.

If you recall, CHP used a set of these Vortec heads on the Goodwrench Quest 350 engine we tested in Chapters 2 through 8. In Chapter 6, Part V, we outlined how we made 408 hp at 5,800 and 430 lb-ft of torque at 3,700 rpm with pocket-ported Vortec heads and 9.5:1 squeeze on the Goodwrench short-block. The rest of the combination consisted of a Comp Cams Xtreme Energy 268 camshaft

One machine operation you should perform is to install screw-in studs. GM Performance Parts studs (right) do not require a pushrod guide plate. The ARP studs (left) will require guide plates. If you use guide plates, do not use rail-type rocker arms. With the GM screw-in studs, no guide plates are required and then rail-type rockers should be used.

(224/230 @ 0.050 with 0.477/0.480-inch lift and a 110-degree lobe-separation angle), an Edelbrock RPM intake, a Holley 750-cfm double-pumper carburetor, and 1 5/8-inch Hooker headers. The test was run on 92-octane fuel but required only 34 degrees of total timing on the dyno to make best power.

As you can see, there's plenty of power potential in these Vortec heads. All you have to do is pick the right parts and the power is there for the taking.

Bolting On Vortec Heads

Edelbrock offers the greatest selection of intake manifolds for the Vortec heads including the new Performer RPM Air Gap intake. They also offer three other heads including a Super Victor single-plane intake.

These GM Performance Parts rail-type roller rockers are an excellent addition to the Vortec head. The rails position the roller tip on the valve stem. This eliminates the need for pushrod guide plates.

From left to right are rail-type roller rockers, stock rail–type rocker arms, and an original nonrail-stamped steel rocker arm. When using any performance camshaft, you will probably need to open up the pushrod holes to ensure the pushrod does not bind in the head.

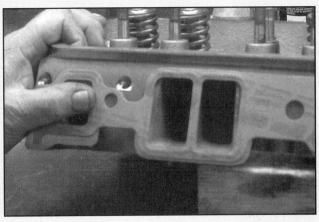

We placed this stock Chevy intake gasket over the Vortec head to show the difference in bolt-hole placement between the two intake manifold bolt patterns.

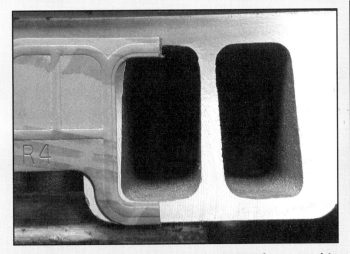

Also note how much taller the Vortec intake port is compared to stock.

Our buddy Tim Moore is building a 355ci street motor to use Vortec heads along with a set of Federal-Mogul hypereutectic pistons (PN 42330) with a slight 12cc dish. Note the slick skirt coating used on the hypereutectic pistons.

Small-Block Chevy Engine Buildups

Compression Lesson

The Vortec head uses a relatively small 64cc combustion chamber. This can cause problems with a typical 350ci small-block with compression ratios in excess of 9.5:1, which can cause detonation problems even with 92-octane gasoline. The example we'll use is a 355ci small-block with 4.030-inch bore, 3.48-inch stroke, and the Vortec's 64cc chamber. The variables we can work with are the piston, deck height, and head-gasket thickness. Basically, changing the deck height by 0.010 inch is worth 0.25 ratio change in compression. The following examples illustrate where the compression falls. For this chart, the 6cc piston is a flat-top piston with four valve reliefs. The 20cc piston is a dished piston. When experimenting with compression-ratio combinations, you need to maintain at least a 0.035-inch piston-to-head clearance. That would be a zero deck with a 0.035-inch-thick head gasket.

Deck Height:	0.020	0.010	0.020	0.020	0.010	0.000
Gasket:	0.038	0.038	0.051	0.041	0.041	0.041
Piston:	6 cc	6 cc	6 cc	20 cc	20 cc	20 cc
Comp Ratio:	9.75:1	9.97:1	9.53:1	8.5:1	8.64:1	8.81:1

Flow Numbers

To put this cylinder head into perspective, we've also included the airflow numbers generated in Chapter 15, "Flow to Go." We've also included flow numbers on a stock 441 iron small-block cylinder head for comparison.

Valve Lift (inches)	Vortec Intake	Vortec Exhaust w/pipe	Iron 441 Intake	Iron 441 Exhaust w/pipe
0.050	40	25	30	20
0.100	70	49	55	45
0.200	139	105	110	83
0.300	190	137	160	118
0.400	227	151	194	133
0.500	239	160	201	146
0.600	229	162	203	148

Parts List

Component	Part Number	Manufacturer
Cylinder head, complete	12558060	GMPP
Cylinder head, bare	12529093	GMPP
Adapters, valve cover	24502540	GMPP
Valvespring kit, ZZ4	12495494	GMPP
Retainer kit, ZZ4	12495492	GMPP
Rocker arm, roller	12370839	GMPP
Rocker arm studs	12371058	GMPP
Intake, dual plane, high-rise	12366573	GMPP
Intake, dual plane, EGR	12496820	GMPP
Intake, single plane	12496822	GMPP
Intake, Performer	2116	Edelbrock
Intake, Performer RPM	7116	Edelbrock
Intake, Performer RPM Air Gap	7516	Edelbrock
Intake, Super Victor	2913	Edelbrock
Intake gasket	12529094	GMPP
Intake bolts	12550027	GMPP
Valve cover, chrome	12355350	GMPP
Valve cover, chrome	141-107	Proform
Valve cover gasket	10046089	GMPP
Head bolts	134-3601	ARP
Rocker studs	134-7101	ARP
Head gasket, 0.038	1003	Fel-Pro
Intake gasket	1255	Fel-Pro

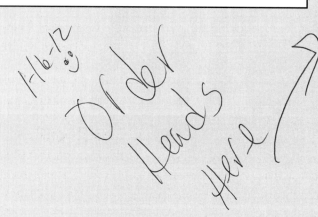

A Tale Of Torque

Dyno Testing AFR's Aluminum 180cc Heads
By Jeff Smith
Photography by Ed Taylor

20

The small-block world just keeps getting better. Earlier we brought you Part V of CHP's "Flow to Go" cylinder-head flow-test series where one of the heads we tested was Air Flow Research's (AFR) new 180cc small-block Chevy cylinder head. On the flow bench, AFR's little intake port outflows many larger-port cylinder heads. Add to this equation the fact that small intake ports create a very responsive engine at low and midrange engine speeds and you have an excellent foundation for a great street small-block. That's exactly why we had to dyno-test these heads.

We already knew from flogging our Goodwrench Quest 350ci small-block that the combination of a good cylinder head (like the GM Vortec iron head and others) along with the right cam, induction, and exhaust pieces is usually worth around 400 hp and 420 lb-ft of torque. With that knowledge base firmly established, we decided to take a typical forged piston 355ci small-block and flog it to see what kind of power we could come up with.

Talk to any good engine builder and he'll tell you that given two sets of heads that measure the same on a flow bench, you should always choose the head with the smaller intake port. This creates greater intake flow velocity and is usually worth additional torque. Since street engines operate over such a wide rpm band from idle to 6,500 rpm or more, greater flow velocities in the intake port contribute to crisper throttle response, more torque, and improved driveability.

The interesting point is that while the rest of the cylinder head world is building larger heads, AFR chose to build a smaller intake port head for the street set. Even more ironic is the fact that this smaller head actually outflows every other head we've tested in the under-180cc category including the Vortec iron head. This head will even outflow many heads in the 180 to 199cc category. Based on these flow numbers, we couldn't wait to put these heads on an engine.

THE ENGINE

To round up a short-block, we contacted Hye Tech Performance and dialed in a 350ci four-bolt main short-block with a CAT steel crank and stock shot-peened rods. The short-

Small-Block Chevy Engine Buildups

Cam Specs

Comp Cams Xtreme Energy 268
Duration: 224/230 degrees @ 0.050 tappet lift
Lift: 0.477/0.480 inch with 1.5 rocker
Lobe Sep. Angle: 110 degrees

Parts List

Component	Company	PN
Block	Hye Tech	stock four-bolt
Crank, 5140 forged	CAT	n/a
Rods, shot-peened	Hye Tech	polished
Pistons, forged	Speed-Pro	LW2603F
Rings, moly (0.030)	Speed-Pro	R9902
Bearings, rod	Clevite 77	CB663
Bearings, main	Clevite 77	MS909P
Bolts, rod	ARP	134-6401
Studs, mains	ARP	134-5401
Oil pan pickup	Moroso	20191
Pickup, for pan	Moroso	24170
Balancer	Summit	SUM-B64262
Timing chain cover	Summit	SUM-G6300
Pointer	Tavia	TAV-02344
Camshaft and kit	Comp Cams	K-12-268-4
Intake	Edelbrock	7101
Gasket, intake	Fel-Pro	1256
Carb, 750 cfm	Holley	0-4779-S
Fuel inlet line	Holley	34-39
Distributor, small cap	MSD	8360
Coil, Blaster II	MSD	8207
Plug wires, universal	MSD	3120
Header, Elite 1 5/8	Hedman	car-specific

block came assembled with a set of Speed-Pro's latest dished D-cup pistons along with Speed-Pro rings and bearings. To this rotating assembly we added a Mellings standard oil pump enclosed with a Moroso kick-out oil pan. After Ed Taylor of Ventura Motorsports bolted on the AFR heads using Fel-Pro gaskets, he also added a Comp 268 Xtreme Energy cam along with an Edelbrock Performer RPM dual-plane Air Gap intake manifold and a 750-cfm Holley double-pumper carb. Finally, we routed the exhaust using a set of 1 5/8-inch Hedman Elite headers.

THE TEST

Taylor took the assembled small-block over to Duttweiler Performance and bolted it to the dyno along with an aluminum Edelbrock water pump and a set of Zoops billet aluminum pulleys. After a quick break-in period, we hit wide open throttle and let this Mouse spin the dial on the dyno. Taylor went through the requisite timing and jetting exercises, but it turned out that the stock Holley carb's jetting was extremely close to perfect while timing ended up at the requisite 36 degrees BTDC.

Once all the tuning was complete, the numbers came out at 435 lb-ft of torque at 4,300 rpm with 416 hp at 6,000. What's impressive about this engine is the torque that comes from the smaller intake port. If you study the dyno chart, you'll see that the 355 cranks out over 400 lb-ft of torque from 2,900 up through 5,400 rpm. It's torque that accelerates the car and this is where this combination really shines. It's actually difficult to clearly identify a peak since the motor makes over 430 lb-ft from 3,600 through 4,600 rpm. That's stout.

We simulated this power curve into a 3,650-pound Chevelle with a 350 trans, 2,100-rpm converter, and a 3.30 rear gear. With good tires at sea level, this motor could easily knock out 12.80s at 107 mph. That's an impressive e.t. and speed for such a heavy car. One reason for this is the strong torque that accelerates this car even without a deep gear.

CONCLUSION

Power is relative. These smaller heads offer a strong torque advantage as well as crisp driveability that would not be possible with larger cylinder heads. These heads would be a great choice for a 350ci or smaller Mouse motor mainly intended for street use. If there is a down side to these heads, they tend to be a bit more expensive than others. But with the impressive flow numbers, this may be a case where spending a bit more money could be a good thing.

A Tale of Torque

We started with a Hye Tech short-block fitted with Speed-Pro forged pistons and steel-forged 5140 CAT crankshaft. The block was bored 0.030-over, torque-plate honed, and also align honed. The forged pistons employ a 12cc dish to keep the compression in line with the 68cc chamber AFR heads.

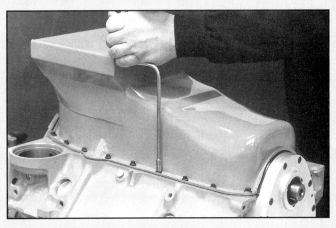

Rather than a stock pan, Taylor opted for a Moroso 7-quart kick-out pan and a dedicated oil pump. Everything was assembled using ARP fasteners.

The induction duties were handled with an Edelbrock Performer RPM Air Gap intake and a 0-4779 Holley 750-cfm double-pump carburetor. The fuel line is also a Holley item that can be adjusted to fit either 4150- or 4160–style Holley carbs.

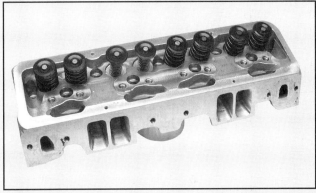

The whole reason for this test was to see how the Air Flow Research 180cc heads would perform. As you can see from the flow numbers we've included, these heads rock. You have a choice of a 68 or 74cc chamber, but all heads come with stainless 2.02/1.60-inch valves, AFR studs and guideplates, and high-quality 1.450-inch valvesprings.

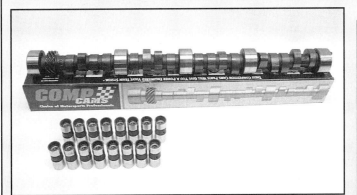

Since we've had such great success with the Comp Cams Xtreme Energy 268 cam, we decided to use it for this test as well. We didn't have time for a cam comparison, but with the AFR heads' excellent exhaust-flow numbers, it would have been interesting to see how a single-pattern version of the 268 would have performed.

With everything bolted up, Taylor set the valves on the hydraulic cam using a set of Comp Magnum roller-tipped 1.5:1 ratio rocker arms. The best setting for most street hydraulic cams is at zero lash plus a half-turn.

Small-Block Chevy Engine Buildups

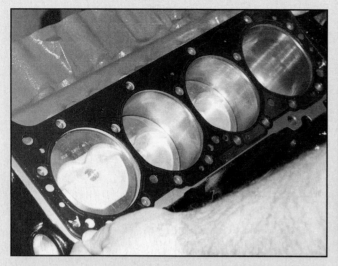

Taylor also used Fel-Pro Perma-Torque head gaskets that employ a solid steel core and a wire ring covered with stainless steel armor. The compressed thickness of 0.041-inch and bore volume of 9.1cc kept the compression ratio within our 9:1 guideline.

To make sure everything stayed cool, we chose an Edelbrock long-style aluminum water pump and a set of Zoops billet pulleys. Behind the crank pulley is a Summit 6.6-inch damper and a Tavia pointer.

FLOW CHART

The beauty of the Air Flow Research 180cc heads is not just the impressive airflow, but also the fact that these are relatively small heads. With only a 180cc intake port, this is only 10 to 15cc larger than a stock set of heads. We've listed a set of stock 882 production iron heads as a comparison. As you can see, the AFR kicks butt throughout the entire curve. The E/I line at the bottom of the chart is the exhaust-to-intake flow relationship expressed as a percentage. Numbers over 75 percent are considered excellent.

Valve lift	Stock Iron 882 Intake	Exhaust w/pipe	AFR 180 Intake	Exhaust w/pipe
0.050	39	34	35	25
0.100	70	59	66	55
0.200	125	109	137	115
0.300	175	136	193	164
0.400	204	143	230	193
0.500	205	144	250	208
0.600	206	145	255	214
E/I @ 0.400		70%		84%

TEST PATTERN

The following is the dyno curve from the 355ci small-block fitted with the AFR 180cc aluminum cylinder heads. The test was performed at Duttweiler Performance and corrected to 60-degree dry air with a 29.95 barometric pressure. Brake-specific fuel consumption numbers—an indication of how well the engine converts fuel to horsepower—average in the low 0.42s. These are very good numbers and indicate an efficient combustion chamber.

rpm	lb-ft	hp
2,600	386	191
2,800	396	211
3,000	408	233
3,200	417	254
3,400	427	276
3,600	430	295
3,800	434	314
4,000	434	330
4,200	434	347
4,400	435	364
4,600	430	377
4,800	421	384
5,000	416	396
5,200	411	407
5,400	400	412
5,600	385	410
5,800	372	410
6,000	364	416
6,200	356	413

Porting for Power

21

A Simple Do-It-Yourself Porting Job
By Scott Crouse
Photography by Scott Crouse

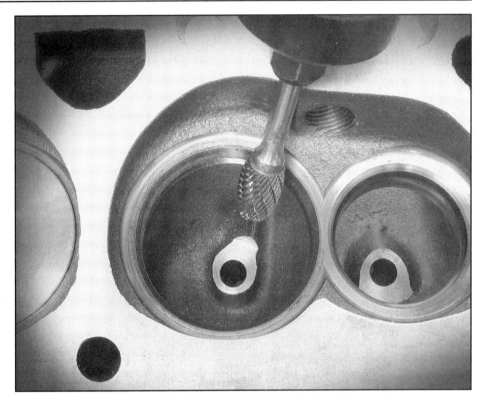

The key to horsepower heaven is in the shape of the cylinder head. Inside these castings, ports direct airflow to and from each cylinder. Since horsepower enthusiasts often attempt to improve their own heads, Chevy High Performance has decided that the time has come to find out what it takes to pull off a successful home-grown port job.

Our first order of business called for selecting a set of heads that still resembled a stock casting. Choosing a pair of high-flow race heads would leave little room for improvement, so we decided on a set of affordable World Products S/R (stock replacement) Torquer iron cylinder heads. World Products offers the S/R Torquer heads with 2.02/1.60-inch intake and exhaust valves in either 64cc or 76cc combustion chambers. While these valves are larger than most 1.94/1.50-inch stock production pieces, we felt they would offer the best potential for our project.

In this chapter, we'll show you how we ported the heads and how the improvements registered on the flow bench. The box-stock S/R Torquer heads flowed a peak 219 cfm on the intake at 0.500-inch lift while the exhaust flowed a max of 134 cfm at 0.550-inch lift (28 in Hg). These are comparable numbers to a stock production iron head.

With the baseline numbers out of the way, we were ready to do some porting. We dove into our Craftsman tool catalog and found a Craftsman Professional 1/4 hp rotary grinder capable of delivering no-load speeds of 26,500 rpm. Standard Abrasives also helped by offering us the use of several polishing kits. These kits work wonderfully for polishing the chambers and exhaust ports as well as deburring a block or removing gaskets. To speed up the cutting procedure, we purchased an egg-shaped carbide cutter with a 1/2-inch head and a 2-inch shank. The plan was to have the author, with no prior porting experience, do the porting to illustrate the ease with which this could be accomplished. What you see here is exactly what the author created on his first attempt.

Before we went to work on our heads we decided it would be best to contact Todd McKenzie of McKenzie Racing in Oxnard, California, for some guidance. From the start, McKenzie made it clear that the goal is to create a smooth-flowing path for the air to travel across. Cylinder head porting is more about how well air flows through the port rather than port size. When we began to remove metal from the bowls it became obvious why it's a good idea to first practice on a junk head. The cutter jumped all

Small-Block Chevy Engine Buildups

Racing flow-tested our World Products S/R Torquer heads before and after the porting procedure to measure our results.

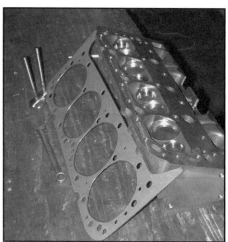

The World Products S/R Torquer cylinder heads are a great performance piece for the enthusiast looking for an inexpensive head that will unleash a few extra ponies. We chose this particular head for its stock-like flow. A higher-flowing head like World's Motown 220 offers much better out-of-the-box performance but would probably be less responsive to a backyard head-porting job.

Before starting our porting adventure, we picked up a Craftsman 1/4 hp rotary grinder and Standard Abrasives supplied a cylinder-head porting kit.

McKenzie showed us how to roll the cutter with the direction of the port while removing material in a 180-degree area.

around and nearly ripped itself from our hands. McKenzie told us that our cutter would work best with a speed reducer, limiting operating speeds to 15,000 rpm. As we began porting, we realized that the new carbide cutter worked extremely fast, requiring us to constantly keep it moving under very light pressure.

Inside the intake and exhaust bowls there's a ridge that forms after the valve seat is machined. McKenzie instructed us to remove this ridge and blend the area in with the contour of the port. With the head placed on its exhaust ports, we began removing material from the ridge in the middle and continued to work our way across the floor and up both walls covering a 180-degree sweep. McKenzie instructed us to remove material only within 1/2 inch of the bottom cut of the valve job. After removing a little material, we used our finger to feel the contour of the port for high spots.

A black marker is handy to have available to mark the high areas. According to McKenzie, the bowls should resemble a venturi where the top of the seat is open and necks down to meet the port where it begins to open back up. Once we removed 180 degrees of the ridge in all the bowls with the head lying on its exhaust ports, we flipped the head onto its intake ports to remove the other 180 degrees of the ridge. When we finished, all of the bowls featured

Porting for Power

The ridge below the valve seat is the key area to remove in order to enhance airflow. Eliminating this ridge develops a smooth radius from the valve seat into the port.

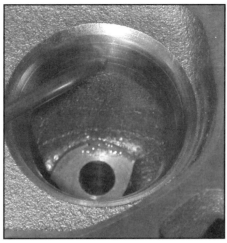

McKenzie instructed us to remove material only within 1/2 inch of the bottom cut of the valve job. This is as close as you want to get with your carbide cutter. The area can later be blended in to form a smooth radius.

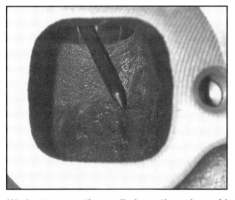

We kept our porting well above the valve guides, but if you're feeling adventurous, it's possible to remove material from this area to pick up a few extra cfm. Be sure to leave a minimum of 0.125 inch of material around the valve guide.

a smooth-flowing 360-degree contour. The important point here is that when porting heads you want to remove the least amount of material possible. Creating concave areas within the port disturbs the air and hurts flow. The entire port should feature a continuous smooth radius. This holds true whether you are porting a small- or big-block cylinder head.

After porting the intake and exhaust bowls, we moved on to the chambers to unshroud the valves. This step involved matching the head gasket with the bolt and dowel-pin holes in order to scribe a cylinder-bore line around the chambers. With a line around each chamber, we could see just how much material could be removed. McKenzie informed us that removing too much material would create serious sealing problems and that it's critical to stay within the line. He also told us that the removed material from the chamber shouldn't increase the combustion-chamber size by more than one or two cc's. Before we went to work on the chambers, McKenzie gave us an old intake and exhaust valve to protect the seats. You can ask your local machine shop for used/damaged valves. Luckily we had the old valves inserted, because we nicked them several times. Having the valves in place won't allow you to touch up the bottom of the chamber wall quite as far down, but it's important to use them. We started our project hoping to avoid the cost of a valve job, but after several nicks during our bowl work it became necessary.

With the bowls ported and the chambers cleaned up, McKenzie instructed us to swap our carbide cutter and use one of the cylinder-shaped polishing rolls from our Standard Abrasives porting kit. Polishing not only creates a smooth path to promote airflow, but also resists carbon buildup. The intake bowls are best left unpolished because it promotes atomization of the fuel. Minor cutter marks within the intake bowls are enough to keep the intake charge tumbling through the ports and into the cylinders.

The last thing we did to our cylinder heads involved port matching the intake runners. In McKenzie's experience, this typically gains an insignificant amount of airflow. The most common type of port matching is done using the intake gasket as a template. Simply scribe a line following the gasket and open the port to match. Be careful in the corners as it's easy to remove too much material with an egg-shaped cutter. It doesn't hurt airflow to leave a little extra material in these areas.

With the porting work complete, McKenzie flowed the heads on his Superflow 600 bench and reported a gain of 11 cfm on the intake at 0.450-inch lift and 7 cfm at 0.500-inch lift. On the exhaust side we saw an outstanding gain of 43 cfm at 0.500-inch lift. The intake gains seemed low

Breathing Easier

Flow test comparison between the stock and ported S/R Torquer heads. All values are in cfm.

Lift	Stock Intake	Ported Intake	Change	Stock Exhaust	Ported Exhaust	Change
0.100	62	59	3	53	56	3
0.150	94	94	–	84	94	10
0.200	127	130	3	105	117	12
0.250	159	161	2	117	148	31
0.300	179	186	7	125	164	39
0.350	196	208	12	128	170	42
0.400	208	221	13	131	172	41
0.450	216	227	11	133	175	42
0.500	219	226	7	133	176	43
0.550	217	221	4	134	178	44

compared to the exhaust. McKenzie informed us that the ridge we removed on the intake side of the port was proportionately smaller than the exhaust-port restriction. Overall, we were very pleased with the flow results, but the only numbers that really matter are torque and horsepower. Next month we'll find out how well we did. We can tell you that the gain was more than 20 hp.

What We Learned

We spent 10 porting hours to gain a maximum 13 cfm on the intake and 44 cfm on the exhaust. Had we attempted to port these heads without McKenzie's guidance we most likely would have cut into the valve seats, hogged out too much material, and hurt the flow on both the intake and exhaust sides. Opening the chambers would have been a disaster without the old valves to protect the seats so we're lucky we had McKenzie available to steer us clear of all these danger areas. The gains were certainly worth the effort.

Be Prepared

When porting heads it is nice to have an air grinder available for variable-speed cutting. Constant use, however, can really challenge even large compressors. During our porting adventure we used the Craftsman rotary grinder most of the time and McKenzie's air grinder for minor touch-ups. The electric rotary grinding tool requires fewer resources but can be difficult to manage without a variable-speed controller. It's also a requirement to wear earplugs, a dust mask, and eye protection. You should also work in a controlled area as metal shavings will go everywhere.

Porting for Power

Before attempting to unshroud the valves, it is important to line up the head gasket with the bolt holes and dowel pins in order to scribe a cylinder-bore line around the chambers.

With a line scribed around each chamber, it is easy to see the amount of material that needs to be removed in order to properly unshroud the valve. It is crucial to not remove material past the line or the head may be ruined.

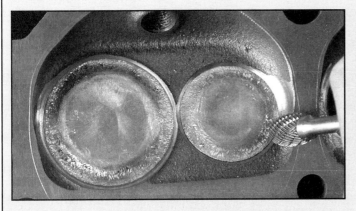

McKenzie lent us some old valves to protect the seats before we started removing metal from the chambers. As you can see, we nicked the valves several times. Be sure to clean the guides with solvent and a pipe cleaner prior to sliding the valves through the guides.

Gasket-matching the intake ports is an easy task, but it's only a small step in moving air through a head. Porting the walls of the runners is time consuming and offers almost no flow improvements.

It is a good thing we contacted Todd McKenzie of McKenzie Racing before we started porting. With his help, we discovered the right way to port our brand-new cylinder heads. Several common porting mistakes include removing too much material, creating concave pits, improper shrouding, polished intake bowls, and ridged contours throughout the ports.

22 Compression Lessons
There's More to Compression Than Just Squeeze
By Jeff Smith
Photography by Ed Taylor

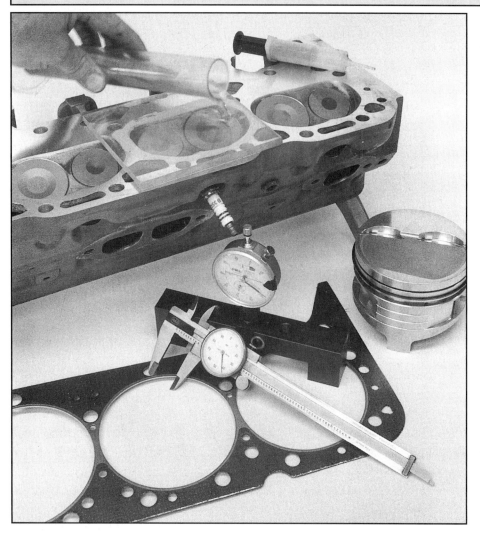

Performance engines are built to push the envelope. A bigger cam, a single-plane intake manifold, a monster carburetor, and large-tube headers all contribute to making more power. Everyone also wants to push the compression as well. Certainly, the harder you squeeze what's between the combustion chamber and the top of the piston, the more power you'll make. But compression isn't always the big power guarantee that many think it is. Let's take a look at compression, how to figure it, and how it plays into the big picture power scheme.

Let's start out by defining compression ratio, which is the ratio of the cylinder volume with the piston at the bottom of its stroke (bottom dead center or BDC) compared to the cylinder volume when the piston reaches the top of its stroke (top dead center or TDC). If we measured a small-block with a cylinder volume of 45ci at BDC and a cylinder volume of 5ci at TDC, then 45 divided by 5 equals 9, giving us a compression ratio of 9:1.

While this sounds simple, it takes accurate measurement and some patience to come up with these values. We'll go over how to measure these different volumes and then show you an easy way to calculate compression.

There are several volumes you must measure to determine compression. The largest is the volume of the cylinder. The other volumes that affect compression include the combustion-chamber volume, the deck height (the height of the piston above or below the block deck), the head-gasket volume, and the piston design that will either add volume (a dished piston) or subtract volume (a domed piston) from the combustion chamber.

All of these components have a direct affect on compression. For example, as the bore becomes larger, compression increases. If you use heads with a smaller combustion chamber, the compression will increase. Even changing to a thinner head gasket will increase the compression ratio.

Compression Lessons

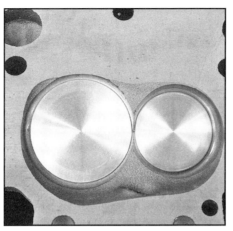

Think all small-block Chevy heads are the same? Think again. With so many different combustion-chamber shapes and sizes, you must measure to really determine an accurate volume. Valves can also have an effect. Notice how the valves in the top photo have a small recess, while the valves in the bottom photo are flat. This will affect chamber volume.

The right way to measure combustion-chamber volume is to invert the head on a workbench and install a pair of valves and a spark plug in the chamber. Use a piece of acrylic plastic that will more than cover the size of the chamber with a 1/4- or 3/8-inch hole drilled in it. Then you can use a colored liquid (blue food coloring in rubbing alcohol works well) to fill the chamber from a graduated cylinder or burette. Powerhouse sells an inexpensive kit using a plastic graduated cylinder that works if you're careful. The 100-milliliter burette is more accurate (1 milliliter = 1 cubic centimeter) but is much more expensive.

Number Crunching

Let's take a look at how all these volumes interact by detailing the simple math that's used to determine compression ratio. The equation for the volume of a cylinder is π (i.e., 3.1417)/4 x Bore x Bore x Stroke. This can be shortened to 0.7854 x B x B x S = volume of a cylinder. For a standard-bore 350ci engine, this computes to 0.7854 x 4 x 4 x 3.48 = 43.73 ci. Since most of the measurements we will be dealing with in the engine are measured in cubic centimeters (cc), we will convert cubic inches into cubic centimeters by multiplying ci times 16.39, which in this case is 43.73 x 16.39 = 716.7 cc.

Rather than try to compute the volume of a complex combustion-chamber shape, it's best to measure the volume with a burette graduated in cc's. For our example, let's say our small-block head comes out to 76 cc. Next, we'll need to measure the piston's deck height. The best way to do this uses a dial indicator on the piston of an assembled short-block. Virtually all small- and big-block Chevy deck heights will place the piston below the deck surface. Often, production engines can have as much as 0.020 to 0.040 inch of clearance between the top of the piston and the deck. Machining the block surface can reduce this distance, an operation referred to as decking. Reducing this distance increases the compression ratio. This can be computed exactly the same way as the cylinder volume, except substituting the deck height for the stroke. Using our 350 example, let's say we've measured the deck height at 0.010 inch. This computes to 0.7854 x 4 x 4 x 0.010 = 0.1256 ci x 16.39 = 2.06 cc.

Piston shape also plays a big role in compression. A pure flat-top piston does not affect compression—but rarely occurs in the real world. Even a "flat-top" piston incorporates valve reliefs in the piston top that add volume. A typical small-block, four-eyebrow piston contributes roughly 4 to 6 cc's worth of volume. This is the same as adding that amount to the combustion chamber size.

Dished pistons perform the same function by increasing the volume above the piston, adding between 10 to 20 cc of volume that reduces compression. Conversely, a domed piston increases compression by displacing volume from the combustion chamber. For example, a piston with a 10cc dome is effectively the same thing as a pure flat-top piston with a 10cc-smaller combustion chamber.

Finally, head gaskets play a critical role in compression ratio and offer the easiest and least expensive route to changing compression. However, head gaskets can be a bit deceiving. You might think that all you have to do to compute the volume is to treat the gasket like deck height. If we compute volume for a standard 350 Chevy gasket like the Fel-Pro 1003 with a compressed thickness of 0.041 inch, we come up with 0.515ci, which equals 8.44cc. But this assumes the gasket bore is both round and the same size as the cylinder bore. In

Small-Block Chevy Engine Buildups

That same plastic plate can be used to check the actual piston and crevice volume. Use a depth mic or dial indicator to determine the piston depth below deck and then, following the example we described in this chapter, you can determine total piston and crevice volume. All this leads to more accurate compression ratio figures.

Any time you bore or add stroke to an engine, the compression ratio will increase. For example, with everything else remaining the same, building a 383 from a standard-bore 350 changes the compression from 8.6:1 to a stout 9.3:1. Boring the cylinders 0.030-inch increased the compression by only 0.10, but the stroke from 3.48 to 3.75 bumps the compression by 0.60.

reality, the 1003 gasket is 4.166 inches in diameter and is not round. The better way to determine compression is to use the manufacturer's published gasket volume. Fel-Pro's published gasket-bore volume for the 1003 gasket is 9.1cc, roughly 0.7cc larger than our computed volume. While this isn't overly critical, it does affect the accuracy of the final result.

Speaking of accuracy, there is another small volume that is usually ignored but also contributes to compression. This is called the crevice volume and is the tiny volume between the compression ring and the top of the piston. Typically this will not increase total volume by more than 1cc, but if you're looking for complete accuracy, it should be included.

The best way to account for crevice volume is to measure the entire piston/cylinder assembly with the engine assembled. Using a flat plastic plate with a hole drilled in the top, you can place the piston 1/2 inch down the cylinder and then fill that volume from a measured burette. Then compare the volume of a perfect cylinder to the amount you measured. The difference will be the combination of any piston top valve reliefs and the crevice volume.

Continuing with our 350 engine example, with a 4-inch bore and the piston top 1/2 inch down the bore, a perfect cylinder would measure 0.7854 x 4 x 4 x 0.50 = 6.28ci x 16.39 conversion = 103cc. Measuring that same cylinder using a four-eyebrow piston resulted in a 110cc volume for a total difference of 7cc. Combining this crevice volume with the more accurate manufacturer's gasket volume produces a much more accurate static compression ratio that can often be 0.10-ratio lower than computed using the less accurate method.

The Easy Way

The sidebar Doin' the Math details the longhand method you can use if you don't have a computer. But for the lazy ones like us with a computer and Internet access, there's a faster way to compute. Performance Trends is a Detroit, Michigan-area company that has been creating excellent high performance automotive software for many years. If you go to the company's Web site (www.performancetrends.com), you can download an easy-to-use compression ratio program for free. The basic program does the math for you instantaneously, which allows you to experiment with all the variables. The program offers several other optional features. If you want the bells and whistles model, it'll cost $40, which is still inexpensive as a time-saver. Just stick it in your hard drive and start calculating.

If you've never computed compression ratio before, this is worth experimenting with on a number of levels. For example, stroke has a much greater effect on compression than bore, which is why the short stroke/small displacement engines

Compression Lessons

Dished pistons like these need to be measured in the bore in order to determine their exact volume. If you measure the piston at its true deck height (0.020 inch for example), you could make the deck height zero in the formula and add the total measured piston dish volume to the chamber volume.

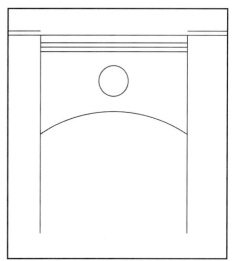

Deck height is the distance of the piston just above or below the block deck surface. The further the piston is below the deck surface, the more volume this adds, which reduces compression. Most engine builders shoot for as close to a zero deck height as possible. Remember that you must maintain a minimum of 0.040- to 0.050-inch of piston-to-head clearance. With a zero deck height, the only clearance is the head-gasket compressed thickness.

Compression Ratio Calculator		(C)1998 Performance Trends Inc.		
Registered to: Jeff Smith				
Engine File: 383CHEV				
Time: 11:23:31:am Date: 13/07/2000				
Report Comment: 383 Small-Block				
File Comment: 383 Small-block with flat-top pistons				

Base Engine Inputs		Calculated Results	Cu. In.	CCs	Liters
Bore, in	4.03	Cylinder Size	47.83	784.	0.784
Stroke, in	3.75				
		Chamber Size	5.84	95.7	0.096
		Compression Ratio	9.19		
Chamber/Piston Inputs					
Chamber CCs in Head	72				
Piston Design: Flat Top w Valve Reliefs		Volume Contributions	Cu. In.	CCs	% of Total
Valve Reliefs, ccs	6	Head Chamber	4.393	72.	75.3
Gasket CCs	9.1	Gasket	0.555	9.1	9.5
		Deck	0.523	8.57	9.
Deck Ht Clearance, in	.041	Valve Reliefs	0.366	6.	6.3
		Piston O.D.	0.	0.	0.

This is what the Performance Trends compression-ratio screen printout looks like. The basic inputs are the same as those used in any compression-ratio program. You can choose from flat-top, flat with reliefs, dished, or domed piston volumes, as well as change the gasket thickness and deck height. The more advanced program does more but you have to buy that one. The outputs on the right are fairly self-explanatory. The volume contributions are expressed in percentages of the total.

require such small combustion chambers. Other areas worth investigating are variables like gasket thickness. Replacing a thick 0.051-inch composition head gasket on a 350 with one of Fel-Pro's 0.024-inch rubber-coated head gaskets can pump the squeeze factor from 8.68:1 to 9.16:1.

Another way to use this program is to play with various volumes to create a suitable combination. Let's say you'd like to use 64cc chamber Vortec iron heads on a 383ci small-block. Unfortunately, this small chamber combined with a flat-top, 6cc valve relief piston, a 0.015-inch deck, and a 0.041-inch thick gasket creates an excessive 10.6:1 ratio—too much for an iron-head street motor on pump gas. Using the software, we discovered that a 20cc dished piston drops the compression to a more 92-octane–friendly 9.2:1 compression.

The speed of this Performance Trends compression ratio program allows you to play tons of what-if games without spending a dime or cursing your calculator. So if learning a few compression lessons sounds like fun to you, log onto Performancetrends.com and download the program. It's fast, it's easy to run, and best of all—you might learn something.

DOIN' THE MATH

If you're stuck on a desert island and need to compute compression, and you left your laptop in your other pair of shorts, here's the longhand version.

In this example, we'll use a standard bore and stroke 350ci small-block (4.00 bore and 3.48-inch stroke) with a 0.041-inch compressed-thickness gasket with a published 9.1cc volume, a 0.010-inch deck height, and a flat-top piston with four valve reliefs with 6cc of volume. The key to working this equation is to know the factor for converting cubic inches to cubic centimeters (ci x 16.39 = cc).

The other way to work this equation is to convert all the cc into ci (cc x 0.06102), but this takes more time. The formula basically divides the total volume of the cylinder with the piston at bottom dead center (BDC) by the volume of the cylinder at top dead center (TDC).

Swept Volume: Bore x Bore x Stroke x 0.7854 = Displacement in cubic inches
                4 x 4 x 3.48 x 0.7854 = 43.73 ci
                43.73 ci x 16.39 = 716.7 cc

Chamber Volume: = Measured 76 cc
Gasket Volume: = Published at 9.1 cc
Deck Height: = Measured at 0.010 inch
                4 x 4 x 0.010 x 0.7854 = 0.1257 ci
                0.1257 ci x 16.39 = 2.05 cc

Piston Volume: = Measured at 6 cc (which includes the crevice volume)

Compression Ratio = $\dfrac{\text{Swept Vol.} + \text{Chamber Vol.} + \text{Gasket Vol.} + \text{Deck Vol.} + \text{Piston Vol.}}{\text{Chamber Vol.} + \text{Gasket Vol.} + \text{Deck Vol.} + \text{Piston Vol.}}$

$= \dfrac{716.7 \text{ cc} + 76 \text{ cc} + 9.1 \text{ cc} + 2.05 \text{ cc} + 6 \text{ cc}}{76 \text{ cc} + 9.1 \text{ cc} + 2.05 \text{ cc} + 6 \text{ cc}}$

$= \dfrac{809.8 \text{ cc}}{93.1 \text{ cc}} = 8.7:1$ Compression Ratio

Part No.	Bore	Compressed Thickness	CCs	Part No.	Bore	Compressed Thickness	CCs
Fel-Pro				**McCord**			
1003	4.166	0.041	9.1	94-2010	4.140	0.042	9.3
1004	4.190	0.041	9.2	94-2011	3.840	0.042	8.0
1010	4.166	0.039	8.9	94-2017	4.190	0.042	9.5
1014	4.200	0.039	9.0				
1034	4.200	0.041	9.3	**ROL**			
1043	4.080	0.039	8.2	HG31010HT	3.825	0.051	9.6
1044	4.200	0.051	11.2	HG31600	4.090	0.020	5.4
1094	4.100	0.015	3.2	HG31000HT	4.175	0.044	11.4
				HG31250HT	4.200	0.051	11.6
Victor Reinz							
3514SG	3.840	0.045	8.4	**Mr. Gasket**			
5746	4.100	0.026	5.4	5727	4.190	0.044	9.8
+1178SSB	4.100	0.020	4.1	5728	4.220	0.042	9.3
+1178SG	4.160	0.045	9.8				
3432SG	4.190	0.045	10.0				
1178SCR	4.190	0.044	9.8				
3512	4.190	0.040	8.8				

New Wave TPI

Scoggin-Dickey's New Intake for The Vortec Head
By Jeff Smith
Photography by Ed Taylor

23

Few cylinder heads have hit the small-block Chevy world with as much impact as the GM Performance Parts Vortec iron cylinder head. By now you already know that the Vortec head offers outstanding intake and exhaust flow combined with an incredible price of under $450 for a complete pair ready to bolt on.

Unfortunately, this cylinder head also brings with it a couple of limitations. First and most importantly, the heads require a specific Vortec intake bolt pattern that differs from all previous small-block Chevy intake patterns. While the Vortec heads can be redrilled for the earlier small-block bolt pattern, it's not recom-mended.

One of the many applications where the iron Vortec head would be a natural is bolted on a third-generation Camaro or Firebird Tuned Port Injection (TPI) engine. Both the 305ci and 350ci TPI engines used iron cylinder heads, and the Vortec would be a perfect swap. Unfortunately, the Vortec's different intake manifold bolt pattern made this swap unrealistic.

Scoggin-Dickey Performance Center in Lubbock, Texas, recognized this opportunity and has just completed a TPI manifold base that employs the Vortec cylinder-head bolt pattern. The manifold base uses larger intake runners to take advantage of the Vortec's excellent intake port flow. The manifold is designed to accept either stock TPI runners or the larger, high-performance runners available from several companies like ACCEL, Edelbrock, Lingenfelter Performance Engineering, TPIS, SLP, and others. These can be used with the stock TPI plenum as well as a stock or large-bore throttle body.

Third-generation Camaros and Firebirds also have emissions legality requirements. The iron Vortec head does not include an exhaust-gas crossover passage, which was incorporated into the third-gen 305ci and 350ci TPI heads. This exhaust crossover port is where the exhaust gas recirculation (EGR) valve picks up exhaust to direct back into the intake. In order for Vortec heads to be legal in a third-generation Camaro, Scoggin-Dickey includes an external exhaust-gas pickup tube that directs the exhaust gas from the passenger-side exhaust header into the intake manifold.

This manifold is not just intended for use on third-gen Camaros however. There's a whole world of opportunities waiting for the chance to pop a TPI-style intake on a hopped-up small-block with Vortec heads. Scoggin-Dickey also sent us several dyno-tests that reveal excellent torque

Small-Block Chevy Engine Buildups

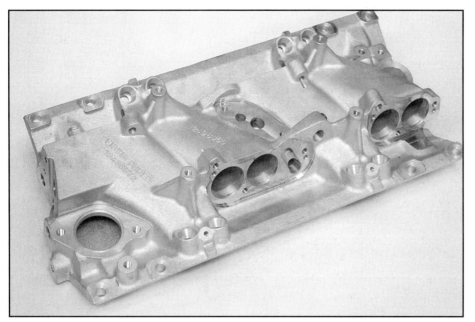

The Scoggin-Dickey manifold allows you to bolt a TPI system on a small-block with Vortec iron heads. The manifold also comes with an external-source tube to plumb exhaust gas into the manifold.

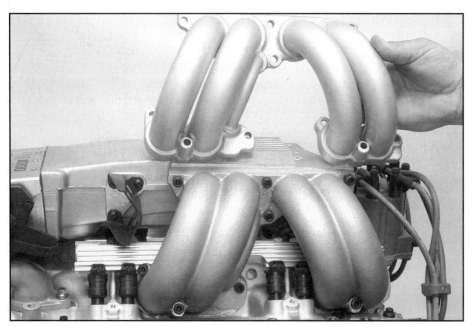

We bolted a set of ACCEL High Flow runners to the Scoggin-Dickey manifold to ensure everything would bolt up. The Scoggin-Dickey intake can accommodate either stock or high-performance runners and plenum.

numbers of over 425 lb-ft at 3,800 rpm and a respectable 354 hp at a very low 4,800. This is a function of the runner length of the manifold that keeps the peak horsepower numbers below 5,000 rpm. While not optimal, this makes for an outstanding grunt motor with tons of torque.

If you're looking for a bit more horsepower, you could also step up to the John Lingenfelter designed ACCEL SuperRam intake. The SuperRam runners are shorter than the TPI runners to increase top-end horsepower while sacrificing little power below peak torque. We've had excellent results with this manifold, and on a 420ci small-block, we've made as much as 525 hp and 540 lb-ft of torque with this intake.

There are a ton of opportunities to combine the excellent dollar-per-horsepower advantages of the iron Vortec heads with this new Scoggin-Dickey Vortec TPI manifold. Whether you want to hop up your third-gen Camaro or just want a slick tip on EFI for your early Chevelle or Camaro, the new Scoggin-Dickey Vortec TPI manifold may be the right part at the right time.

The Art of Installation

We ran into a couple of little catches when bolting the ACCEL Hi-Flow runners to the Scoggin-Dickey intake base. The ACCEL runners needed a slight grinding on the bottom of one corner of each pair in order to clear the Vortec intake bolts, which are in a different place than they are on the original intake.

The key to assembling either a TPI or an ACCEL SuperRam is pre-assembly. Bolt the intake manifold to the heads and torque it in place. Then start all the bolts for the runners and plenum and make sure everything locates properly. This will save time and frustration. Also be aware that the bolt patterns for the runners are identical, but the runners are dedicated left and right due to different EGR passages. This means the runners can be installed backwards, but this will create a vacuum leak. Be sure to match up the runners and gaskets to the plenum to ensure the system is installed correctly.

New Wave TPI

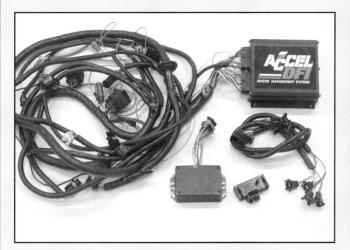
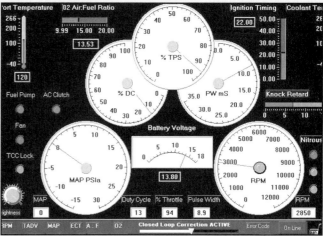

ACCEL also offers a complete stand-alone fuel injection system that has just been entirely redesigned using Windows-based software. The system offers significant advances over the previous system at a reasonable price.

ACCEL has 13 different high-impedance injectors ranging from 14 to 48 lb/hr output. This is the manifold with an ACCEL fuel rail and injectors, but a stock fuel rail will also bolt right up.

One way to increase fuel flow is to increase the fuel pressure. ACCEL offers an adjustable fuel-pressure regulator that bolts to the stock fuel rail that will increase line pressure for tuning purposes.

The Scoggin-Dickey intake also comes with an external EGR tube that connects the exhaust line from the header or exhaust manifold to the intake. The EGR-valve is located in the middle of the intake under the plenum. This external exhaust-gas plumbing makes the Scoggin-Dickey manifold legal for swaps in third-generation Camaros.

John Lingenfelter designed this ACCEL SuperRam to pump up the horsepower on a typical TPI system. The shorter runners allow the engine to extend the peak horsepower up to around 5,800 rpm.

141

Small-Block Chevy Engine Buildups

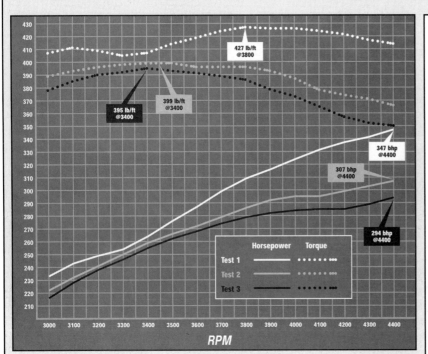

TEST NUMBERS

Rpm	Test 1 TQ	Test 1 HP	Test 2 TQ	Test 2 HP	Test 3 TQ	Test 3 HP
3,000	378	216	389	222	407	233
3,100	385	228	393	232	411	243
3,200	390	238	396	241	409	249
3,300	392	246	398	250	405	254
3,400	395	255	399	259	407	264
3,500	393	262	399	266	414	276
3,600	391	268	396	272	419	287
3,700	389	274	396	279	424	299
3,800	386	279	395	286	427	309
3,900	379	282	393	292	426	316
4,000	373	284	387	295	426	324
4,100	365	285	378	295	424	331
4,200	357	285	374	299	421	337
4,300	352	289	371	303	417	341
4,400	350	294	366	307	414	347
4,500	347	297	362	311	409	350
4,600	345	302	356	312	402	352
4,700	341	305	353	316	394	353
4,800	333	305	346	317	387	354
4,900	323	301	337	315	378	353
5,000	310	295	327	311	370	352
5,100	300	291	315	306	362	351
5,200	292	289	306	303	353	350
5,300	284	287	297	300	346	349
5,400	278	286	290	299	340	350
Avg.	353	278	365	287	400	317

Scoggin-Dickey sent us three interesting dyno-curves based on early testing of the Vortec big-port intake. Test 1 is a stock L98 350ci short-block with iron Vortec heads, a factory L98 hydraulic-roller cam (see the "Cam Specs" chart for details on all the cams), the Scoggin-Dickey base, stock TPI runners, plenum, and throttle body. The exhaust used open 1 7/8-inch headers. Test 2 was the same engine with a ZZ4 hydraulic-roller cam and LT4 valvesprings. Scoggin's Test 3 added a set of Edelbrock runners and throttle body along with a GM Performance Parts HOT cam and 1.6 roller rockers. As you can see from the numbers, the motor really comes alive with the bigger Edelbrock runners, throttle body, HOT cam and rockers. The peaks of 427 lb-ft of torque and 354hp at relatively low engine speeds are strong but by no means optimized. Changing to smaller 1 5/8-inch headers and a good low-restriction dual 2 1/2-inch exhaust system would increase torque dramatically. According to Scoggin-Dickey's Nickey Fowler, they performed this test using stock '87 factory F-car fuel and spark curves with slightly larger 24 lb/hr injectors.

CAM SPECS

Cam	Duration @0.050-inch (degrees)	Valve Lift (inches)	Lobe Separation Angle (degrees)
L98	I 202	0.403	115
Corvette	E 206	0.415	
ZZ4 Hyd.	I 208	0.474	112
Roller	E 221	0.510	
HOT Hyd.	I 218	0.525*	112
Roller	E 228	0.525*	

*1.6:1 roller rockers

It's A Spring Thing

24

How to Change Small-Block Valve Springs
By Jeff Smith
Photography by Jeff Smith

Greg Smith had a problem. His '55 had come up lame, and he wasn't sure what it was. "It just lays down at the top of first gear," he said. The '55 used to be a stab-it-and-steer-it 12-second car with its 355ci small-block, Turbo 400 trans, and 4.33 gears. But lately the car had slowed down considerably and now was running disappointing 13.20s at 106 mph.

At first, friends all chipped in with suggestions. Greg improved the fuel delivery system thinking that the motor wasn't getting sufficient fuel. He added a big electric pump and lines, but the problem persisted. Then he borrowed an MSD-6A ignition box and distributor and while the throttle response improved dramatically, the car's performance was still lame.

Finally, someone suggested checking the valvesprings. This made sense since Greg has always shifted this engine up around 6,800 rpm. While Greg doesn't drive the car on the street much, it has seen its share of dragstrip duties over the years. Greg pulled both valvesprings from the No. one cylinder and trekked down to his local machine shop where they pronounced the springs mortally wounded. The seat pressure that should have been around 100 to 105 pounds was down around 80 pounds, while max lift pressure was proportionately low. Clearly, it was time for a new set of springs.

Swapping valvesprings is an easy job that can be done in your driveway with a few special tools. You'll need an air compressor, hand tools, a valvespring compressor, an adapter to plumb compressed air into the cylinder, and a small pencil magnet. We used part of a compression gauge to plumb air into each cylinder. The air will prevent the valves from dropping down into the cylinder when you remove the retainer.

SPRING DEATH

So what killed these springs? The smoking gun lies with rpm. Everybody likes to buzz the motor to the moon. There's nothing that sounds sweeter than a small-block revving to 8,000 rpm as the driver rows his way through the gears. But this same rpm is absolutely death on valvesprings. In Greg's case, the springs were of questionable origin and had been on the engine for many moons. Subjected to hundreds of rpm blasts both on the street and on the track, the springs finally gave up and lost much of their original pressure.

When the spring can no longer control the valve, the engine goes into what is called valve float. Most enthusiasts think this means the valve launches off the nose of the cam at max lift and can smack the piston. While this can happen, the more common occurrence is the valve bounces off its seat upon returning to the closed position. In the case of the

143

Small-Block Chevy Engine Buildups

intake valve (which is heavier than the exhaust), this bounce contributes to lost cylinder pressure because the valve is off the seat when it should be closed. The less pressure the spring exerts on the valve, the more times it will bounce and the more cylinder pressure is lost. This is why the engine just quits revving. That was Greg's complaint with the car.

THE FIX

The good news is that the fix is easy. All you have to do is choose the proper springs for the cam, and the cam manufacturers have already done that for you. In Greg's case, the cam was a big Crane mechanical flat-tappet cam with 0.554/0.572-inch lift (see "Cam Specs" for the details). The combination requires a stout dual valvespring. We also double-checked to make sure the new spring diameter was the same so no machine work would be necessary.

Once the springs arrived, Greg used a slick valvespring compressor tool from Moroso to remove the springs. The photos reveal how easy this was, but don't be fooled. The job still took the better part of the day to complete, but that was mostly because the '55 Chevy engine compartment was somewhat confining, and also because the headers had to come off in order to install the compressed air attachments.

Greg also took the time to carefully measure each spring's installed height to ensure that the springs would have the proper pressure (see "The Right Height"). This added to the time it took to complete the job, but it's the right way to do it. With the springs installed, Greg bolted the Crane gold roller rockers back on, set the lash to Crane's spec, and fired the motor.

CAM SPECS

Crane 260/3694-2S-6
Mechanical flat-tappet cam

	Duration (advertised)	Duration @.050	Lift (in., 1.51)	Lobe Sep. (deg)	Lash (hot, in.)
Int.	296	260	0.554	106	0.026
Exh.	304	268	0.572	—	0.026

Crane valvespring recommendation:
- PN 99893 dual valvespring 4,500 to 7,500 rpm range
- Closed pressure at installed height: 1.875-inches, 120 pounds
- Open pressure at 0.525-inch valve lift: 323 pounds
- Coil bind @ 1.080 inches
- Max valve lift: 0.795 inch

To revive his aging small-block, Greg replaced the springs, retainers, and keepers with new parts from Crane. Crane lists the recommended spring part number right on the cam card, or you can look up the appropriate parts in the catalog.

After readjusting the valve lash when the temperature came up, he was ready for a quick test run. There wasn't time to go to the dragstrip to test the car, but Greg reports that the car feels much stronger and will now rev right to 6,800 rpm without a problem.

Valvesprings may not be very glamorous, but they are an essential component if you are going to buzz your motor much past 5,000 rpm and expect it to make power up there. Match the springs to the cam and install them properly, and you can expect healthy horsepower dividends.

It's A Spring Thing

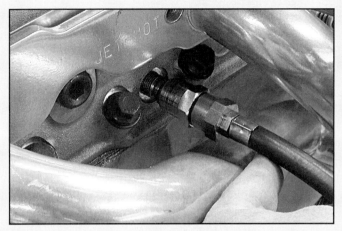

To start, pull the spark plug and the rocker arms for only the cylinder you are working on. Use the adapter to plumb compressed air into the cylinder. This may require loosening or removing the headers from the engine to fit the adapter. The air may turn the engine over slightly, but that's normal.

Use a small hammer to tap each valvespring retainer. You'll actually hear the difference when the keepers loosen up. If you don't do this, the keepers will stick to the retainer and you won't be able to remove them from the retainer. You can use a small pencil-type magnet to retrieve the loose keepers.

With air in the cylinder, Greg used this slick Moroso valvespring tool to compress the spring and remove the keepers and retainer. The tool screws onto the stud and uses compound leverage to compress the spring.

Just pull the handle; the spring compresses, and everything comes apart. The tool even over-centers so you don't have to hold the lever down while you work.

Before installing the new spring, Greg used a height mic to measure the installed height distance between the spring seat in the head and the bottom of the retainer. In Greg's case, the spec was 1.875 inches.

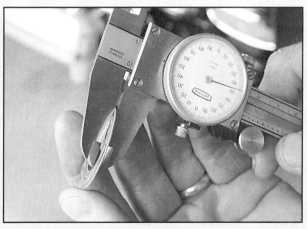

Each spring location is a little different. In this case, the distance measured 1.905 inches, or 0.030 inch more than spec. Greg used a 0.030-inch spring shim that was the same outside diameter as the spring and placed it on the spring seat.

145

Small-Block Chevy Engine Buildups

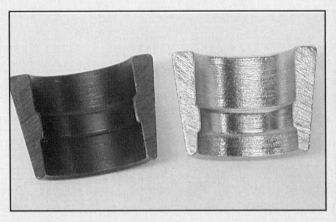

Crane also offers keepers that place the lock groove in three different positions, which will move the retainer up or down on the valve. Crane offers standard, +0.050-inch and –0.050-inch keeper heights that can vary the installed height of the spring using the keepers. This is a very slick idea.

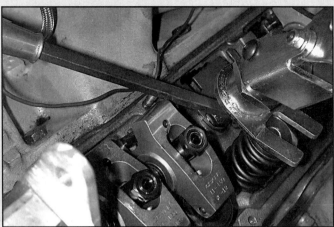

Once the proper installed height is set, you can reassemble the spring on the seat. Sometimes you have to pry on the spring slightly to line up the retainer with the valve to allow installing the keepers.

You might also consider cutting the heads for positive-control valve seals. For small-block Chevys, this requires machining the outside diameter of the guides. This means removing the heads from the engine because you must also remove the valves. However, Crane sells a tool that allows you to machine your heads at home using just a 1/2-inch drill motor.

This illustration shows the relationship between the seal and the retainer, which you should also check. Measure the distance between the bottom of the retainer and the top of the seal. Subtract the valve lift and you should have at least 0.050 inch of clearance. Greg had over 0.100 inch, which is plenty.

To properly install valvesprings, you also have to measure the installed height. All springs increase the load based upon how much the spring is compressed. The installed height is the slightly compressed height of the spring with the valve in its fully closed position. For Greg's engine, Crane specified the installed height at 1.875 inches. This is the distance between the retainer and the spring seat in the head (see illustration A). The most accurate way to measure this is to use a slick little tool called a height mic, which can be purchased from Crane. This inexpensive tool makes measuring installed height very easy. Using the exact retainer and pair of keepers for that valve, slip the height mic over the valve stem, then install the retainer and keepers. When the mic is tight between the spring seat and the retainer, you can read the installed height. If the measured height is more than spec, you can use shims to come within 0.010 inch of the installed height. Now you are assured that the spring is installed with the proper seat tension. If the height is less than spec by less than 0.010 inch, this will still work.

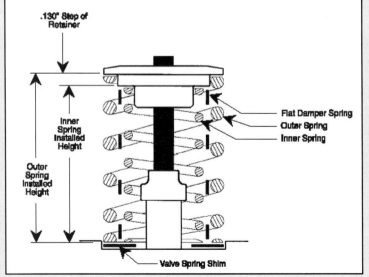

Angling for Power

How a Simple Valve Cut Can Increase Power 20 Percent
By Jeff Smith
Photography by Jeff Smith

25

As all good engine enthusiasts know, there's horsepower to be found in a good set of cylinder heads. But even stock cylinder heads can benefit from seemingly insignificant changes to the valve job. For years, cylinder head experts have known that adding a single angle cut to an intake or exhaust valve can increase low-lift airflow. We decided to test this idea on a stock set of iron small-block cylinder heads to see just how much this small change might be worth. But before we get into the test, let's review a typical valve job to explain how all these crazy angles come together.

Degrees of Separation

Any discussion about a professional valve job must start with tight valve guides and straight valves to ensure accurate placement of the valve on the seat. The basic high-performance valve job consists of three angles: a 45-degree seat angle bracketed by 30- and 60-degree angles. This seat angle matches an identical 45-degree angle machined into the valve. Since air does not like to make abrupt angle changes, stock heads employ a throat angle (below the seat) of roughly 75 degrees and a top angle (above the seat) of 15 degrees. This forms a 15-45-75-degree series of angles that creates a radius that directs air past the valve. For the exhaust side, these angles may change slightly to 60-45-15 degrees.

For street engines, the intake seat width is usually machined to around 0.050 inch wide. Since the exhaust valve operates at much higher temperatures, a slightly wider seat of 0.080 inch is desirable to conduct heat away from the valve to prevent damage.

Another valve-job variable is the placement of the seat on the valve itself. According to cylinder head porter Todd McKenzie, the factory positions the seat exactly in the middle of the valve. But any good cylinder head specialist knows that placing the seat higher on the valve (closer to the outside diameter of the valve) improves flow. Unfortunately, this also sinks the valve in the seat, but the net result is still improved flow. Combine a narrower seat width with a higher placement on the valve and this adds up to increased flow.

Cuts Like a Knife

For this chapter, McKenzie changed the factory valve-seat angles to a more performance-oriented 30-45-60-degree configuration and then tested the same port again. As you can see from the flow test chart on page 150, these changes improved the exhaust-side flow by as much as 17 cfm at

Small-Block Chevy Engine Buildups

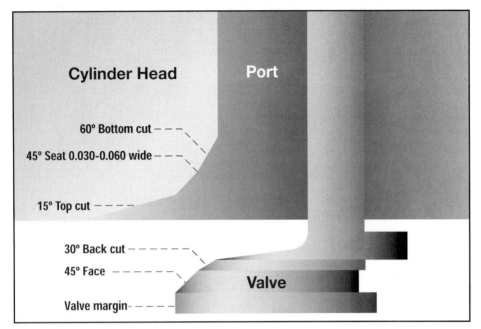

This illustration gives the best view of the relationship between the intake valve and the seat. Note the 45-degree face on the valve and its corresponding 45-degree angle on the seat with the adjacent 30- and 60-degree angles. This shows how the 30-degree back cut improves the radius.

0.50-inch lift (a 13 percent improvement) but were responsible for only minor 1 to 2 percent changes on the intake side.

This set the stage for testing the 30-degree back cut. The key here is the fact that the stock 1.94/1.50-inch stock valves are more restrictive on the back side than performance stainless steel valves. The back cut removes some of this valve material and improves the radius past the valve seat, which improves the flow. McKenzie placed a 30-degree back cut on both the intake and exhaust valves and machined the back cut all the way down to just shy of the back side of the seat. He also added a 23-degree cut inboard of the 30-degree seat on the exhaust valve. This 23-degree cut is machined to produce 30- and 23-degree cuts of the same width.

This created an exhaust valve with three distinct angles of 45-30-23- degrees. The idea behind the 30-degree back cut is to create a radius in the valve as well as the seat in search of increased airflow. As you can see from the flow data, the 30-degree back cut dramatically improves the low-lift flow numbers by 0.350-inch valve lift, yet becomes negligible above 0.350-inch valve lift.

Theoretical Power

It might seem strange to be so infatuated with flow numbers below 0.300 inch, but for a street engine with a typical maximum valve lift of 0.450 inch, the piston's maximum acceleration rate away from top dead center (TDC) occurs right around the valve's 0.300-inch lift number. This is when the cylinder's demand on the induction side is the greatest. Therefore, enhancing a cylinder head's low- and mid-lift flow rates offers outstanding potential for increased power—especially torque. The valve also "resides" at these low-lift flow points for a much longer period of time than it does at maximum valve lift. Keep in mind that the valve is at these low- and mid-lift flow numbers twice during its journey from open to closed while it reaches max lift only once.

For the exhaust side, cylinder pressure is the highest when the exhaust valve first opens. Therefore, any change that improves low-lift exhaust flow should improve cylinder scavenging and reduce pumping losses. These losses tend to occur on engines with restricted exhaust systems that require the engine to expend power to pump the remaining exhaust past the valve and out of the cylinder. This residual exhaust pressure also increases the likelihood of exhaust gas entering the intake tract when the intake valve first opens.

The beauty of back-cutting valves is that the price is more than right. A machine shop will probably charge less than $75 to back cut all 16 valves. While this isn't dirt cheap, you are virtually guaranteed a power return for this investment. McKenzie estimates that a professional valve job with back cuts on a good set of cylinder heads used on a 400hp engine could be worth between 10 and 20 horsepower.

While it may not be worth it to tear your engine apart just to do a valve job and back-cut the valves, this is great information to log into your memory bank when it comes time to build that next great Chevy motor.

Angling for Power

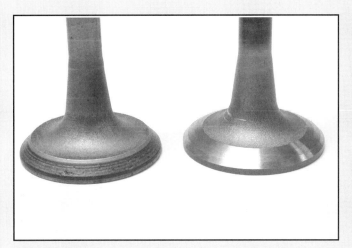

You can see the extra material on the stock exhaust valve (left) compared to the stainless exhaust valve (right). Many high-performance valves eliminate this added material, making a back cut unnecessary.

The width of the 45-degree seat is also critical to airflow. This intake valve has been hand-lapped and indicates a relatively wide seat of over 0.060 inch. The total valve-seat width is only 0.080 inch. Narrowing the seat width to around 0.040 inch would improve airflow.

McKenzie's flow testing reveals just how much a simple angle change is worth.

If you really want to see improvements, combine these valve-angle tricks with a quality pocket-porting job and it's possible to dramatically improve the low- and mid-lift flow numbers on even a stock iron head.

Look closely at this stock exhaust valve and you can see the 45-degree seat angle, the 30-degree back cut, and the faint parting line for the final 23-degree cut.

149

Small-Block Chevy Engine Buildups

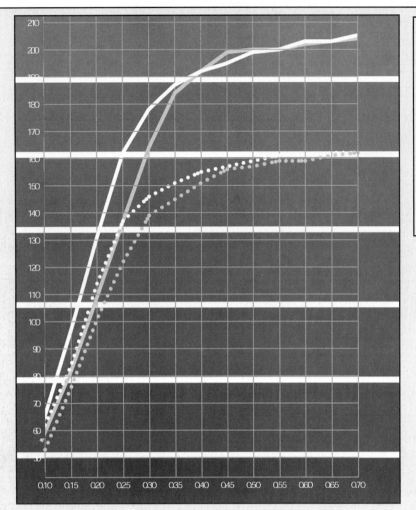

The test head is a stock iron 882 casting with 1.94/1.50-inch stock valves. All tests were performed at McKenzie Cylinder Heads using a SuperFlow 600 flow bench with 28 inches of water test depression and a 4-inch bore. All exhaust-port flow tests used a 1.625-inch flow pipe.

Test 2 Intake
Test 2 Exhaust
Test 3 Intake
Test 3 Exhaust

TEST

Valve Lift	Test 1 I	Test 1 E	Test 2 I	Test 2 E	Test 3 I	Test 3 E	% Increase I	% Increase E
0.100	59	47	59	53	65	62	10	17
0.150	80	65	82	76	97	84	18	10
0.200	109	87	109	101	131	114	20	13
0.250	139	106	137	122	162	137	18	15
0.300	164	125	164	139	178	146	8	5
0.350	180	134	184	145	187	151	1	4
0.400	190	137	192	151	192	155	—	2
0.450	195	139	199	156	195	157	−2	—
0.500	197	140	200	157	199	159	—	1
0.550	199	140	200	159	200	161	—	1
0.600	200	140	202	159	203	161	—	1
0.650	202	140	203	161	203	162	—	—
0.700	203	142	204	162	205	162	—	—

Test 1: The baseline consisted of stock 1.94/1.50-inch valves and a factory valve job. The valve job used 15-45-75-degree intake-seat angles with 15-45-60-degree exhaust-seat angles.

Test 2: McKenzie narrowed the intake-seat width to 0.050 inch and moved the seat to 0.005 inch inside the outer edge of the valve. He also narrowed the exhaust seat to 0.060 inch and moved the seat out to 0.005 inch inside the outer edge of the valve.

Test 3: For this test, McKenzie placed a 30-degree back cut on the intake down to the top of the lap line. He performed the same change to the exhaust valve but added a second, 23-degree back angle to the valve until the 23-degree and 30-degree angles were the same width.

Cam Basics

Understanding Cam Lift and Duration
By Jeff Smith
Photography by Jeff Smith

26

If the cylinder heads represent the aerobic side of an engine, and the rotating assembly is its muscle, then the camshaft must be the brain. The cam is the component that signals when the valves open and close, timing the valve events to create power.

Appropriately, the cam is also the most complex and often baffling component in an internal combustion engine. That lumpy-bumpy stick with all the lobes commands a whole vocabulary of terms that can be especially confusing. This brief primer on camshaft basics will decipher many of those terms and explain how a camshaft works. Once you have a general understanding of what all this cam talk is about, you can use that knowledge to pick the cam that's best for your engine. We can narrow the discussion of how a cam works down to lift, duration, and overlap as the three most critical components of cam design.

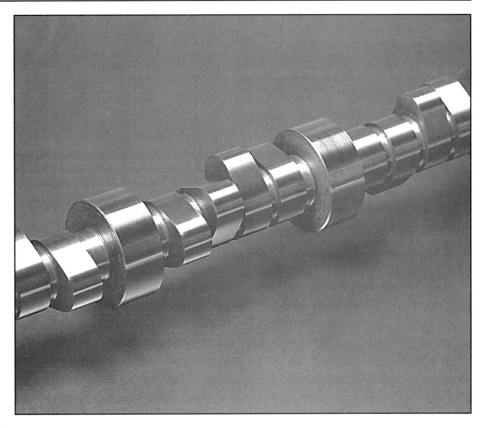

Lift

A camshaft transforms rotating motion into linear, or straight-line, motion. Whoever designed the first camshaft lobe probably had no idea how complex this egg-shaped device would become. All eccentrics are based on a circle. In camshaft terms, this is called the base circle. From there, you add an eccentric, or lobe, that creates lift. The height of the lobe above the circle's radius is the lobe lift of the camshaft.

Let's say we have a lobe lift measuring 0.333 inch. Since the Chevy engines that we're interested in are pushrod V-8s, we also must include a rocker arm. The rocker is actuated by the vertical motion of the lifter and pushrod and multiplies lobe lift by the rocker arm ratio. For the classic small-block, the standard ratio is 1.5:1. Multiplying the lobe lift times the rocker ratio will produce gross valve lift (0.333-inch lobe lift times 1.5:1 rocker ratio = 0.499-inch gross valve lift).

Higher rocker ratios can significantly increase valve lift, but this requires stiffer components that can become rather expensive. However, if you have a cylinder head that can flow more air at greater valve lifts, there is power to be made. Creating more lobe lift is not as easy as just making the eccentric taller. Because of physical limitations between the lifter and the cam, more lobe lift requires more distance to create the opening and closing ramps. This is created by extending the length of the ramps.

Small-Block Chevy Engine Buildups

Camshafts come in either flat-tappet (bottom) or roller (top) versions. Within each of these styles, you can opt for either a solid or hydraulic version of the cam and lifters. Roller cams offer more lift for the same amount of duration compared to flat-tappet cams.

Of the two styles, hydraulic cams are far more popular than solid cams, mainly because solid cams require periodic maintenance to set the valve lash. However, with quality parts like a good poly lock, setting lash is an infrequent requirement.

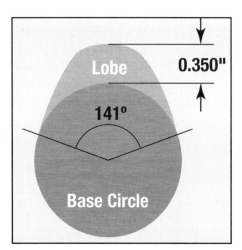

The height of the lobe above the base circle is the amount of lift generated by a cam lobe. Multiply lobe lift times the rocker-arm ratio to establish valve lift. For example, a 0.350-inch lobe times a rocker ratio of 1.5 equals a gross valve lift of 0.525 inch.

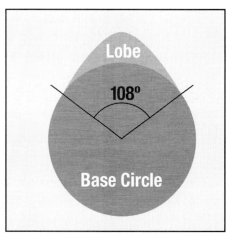

A lobe with 108 degrees of duration at 0.050-inch tappet lift multiplied by 2 (because the cam spins at half the crankshaft speed) equals a duration of 216 degrees, which is a relatively mild lobe.

Duration

Duration is the term given to the amount of crank movement (in crankshaft degrees) that the lobe creates lift by pushing the lifter off the base circle. We'll go over each of these specs to make them easier to understand. If you look at the size of a camshaft gear relative to the crank gear, you'll notice that the cam gear is twice the size of the crank gear. This means that the camshaft spins at half crankshaft speed. To make it easier to understand cam duration, most cam specs are given in crankshaft degrees.

Duration is expressed as the number of degrees of crankshaft rotation used by the lobe to move the lifter. Since the exact point at which the lifter starts to move can be difficult to establish, cam companies use a checking height to establish this movement. The SAE standard is 0.006 inch of lifter rise off the base circle of the cam. Unfortunately, not all companies use this standard. Because different checking heights can make a large difference in the "advertised" duration of a camshaft, the industry established the common checking height of 0.050 inch for duration. This is the only way to accurately compare duration figures of camshafts from different manufacturers.

For each lobe there is an opening and closing point. Duration is expressed as the number of degrees that the crankshaft rotates between these two points. For example, the intake lobe on a Crane PowerMax H-288 cam has an advertised duration of 288

Cam Basics

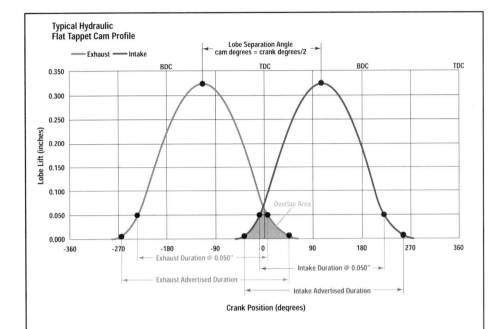

GRAPHIC UNDERSTANDING

This one graph is probably a better tool to understanding camshafts than pages of words. The lobes are represented here with lift in inches expressed on the vertical scale and degrees of duration across the horizontal scale. The red exhaust lobe opens -180 degrees before BDC and closes after TDC (0 degrees). The blue intake lobe opens before TDC (0 degrees) and closes after BDC (180). Note that both valve events require almost one entire rotation of the crankshaft (360 degrees). Keep in mind that the four-stroke cycle requires 720 degrees, so there is plenty of time for the compression and combustion cycles.

Note the green-colored triangle shape that denotes the overlap area. This is the point where the exhaust valve is just closing and the intake valve is just opening. There's much more happening here than it may appear, which is why we've devoted an entire chapter to overlap (see "Cam Overlap Chronicles," Chapter 27.).

Also note the positions of the intake and exhaust centerlines. The number of degrees between these two centerlines is expressed as the lobe separation angle. For example, let's say the exhaust lobe centerline is 117 degrees BTDC and the intake centerline is 107 degrees ATDC, you can add these two values together and divide by two to produce the lobe separation angle (117 + 107 = 224/2 = 112 degrees). This can get confusing since the lobe separation angle is often the same number as the intake centerline, but they define two completely different values. The lobe separation angle is expressed in camshaft degrees while intake centerline is always expressed in crankshaft degrees

degrees and a 0.050-inch duration of 226 degrees. The 0.050-inch duration will always be less than the advertised duration because it measures between opening and closing points on the lobe that are closer together—where the tappet reaches 0.050-inch rise. The advertised duration for Crane cams is measured at 0.004-inch tappet lift, which encompasses a greater rotation, so the number of degrees will always be higher.

We should also discuss where these opening and closing points occur. Intake opening (IO) usually occurs before top dead center (BTDC), while intake closing (IC) happens after bottom dead center (ABDC). For the exhaust side, exhaust opening (EO) occurs before bottom dead center (BBDC) and exhaust closing (EC) after top dead center (ATDC). These data points are listed on the cam card that comes with each new camshaft.

These points can also be verified with the cam in the engine when the camshaft is degreed. If you're not sure of the size of the cam but you know the intake and exhaust opening and closing points, you can determine the duration by simply adding the opening and closing points to 180 degrees. For example, the Crane PowerMax 288 cam's 0.050-inch tappet lift numbers are IO at 4 degrees BTDC and IC at 42 degrees ABDC. Add these to 180 degrees and you get 226 degrees (4 + 42 +180 = 226 degrees). This technique also applies to the exhaust lobe.

Intake Centerline

Now that you're conversant with lift and duration, let's add another detail called the lobe centerline. Imagine looking at a lobe split directly down the middle as viewed from the end of the cam. This line would create what's called the intake centerline. The exhaust lobe also has a similar centerline. If you look at a typical cam card, the intake centerline is also expressed in crankshaft degrees ATDC. For example, a Comp Cams 268 Xtreme Energy cam has an intake centerline of 106 degrees ATDC. You can use this information to degree the cam and find the intake centerline for cylinder No. 1 to ensure the cam is installed in the proper relationship to the crankshaft.

When degreeing the cam, it's possible that the cam may not always check out in the right place. For example, let's say you degree this cam in your engine and discover the intake centerline is actually installed at 112 degrees ATDC compared to the 106 degree specifications. This means the cam is retarded relative to the cam card spec. To position the cam at 106,

153

Small-Block Chevy Engine Buildups

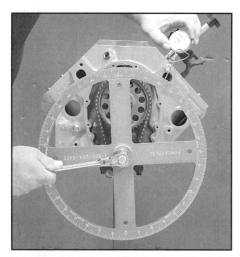

Degreeing a camshaft is not required for installation, but it's good to know exactly where the cam is. Our experience is that cams usually bolt in within 1 to 2 degrees of their intended position, but we've also seen them installed and retarded by as much as 4 degrees.

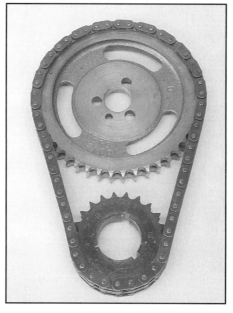

The camshaft spins at exactly half the crankshaft speed because the cam gear is twice the diameter of the crank gear. This timing-gear set offers several keyway slots in the crank gear that allows the engine builder to advance or retard the cam. Another option is offset bushings (inset) that fit over the cam pin to advance or retard the camshaft.

the cam must be advanced 6 degrees. Conversely, if we found the cam at 103 degrees ATDC, this is advanced compared to the 106 centerline. Repositioning the intake centerline at 106 degrees ATDC would require retarding the cam 3 degrees.

Single- and Dual-Pattern Cams

We're slowly assembling the building blocks that make up a camshaft. If you've ever looked at the card that comes with a new cam, then perhaps you've noticed that sometimes the intake and exhaust duration specs are similar and other times they are not. A single-pattern cam uses the same duration and lift numbers for both the intake and exhaust lobes, while a dual-pattern cam often employs a longer duration and more lift on the exhaust side. A dual-pattern cam employs more exhaust duration in order to compensate for a weak exhaust port, which is often the case with stock heads.

An interesting phenomenon is now occurring with regard to single- and dual-pattern cams. Before the days of excellent aftermarket heads for the small-block Chevy, most cam manufacturers built single-pattern cams. Once these companies discovered that dual-pattern cams made more power, these cams became the new hot ticket. The emerging pattern now is that most aftermarket heads offer such strong exhaust ports that the shift to a single-pattern cam will make more power. Expect to see the cam companies again move back to an emphasis on single-pattern cams as cylinder heads continue to improve.

Cam Selection

It's tough to condense this massive subject down into a few short paragraphs, but the key to cam selection is to be brutally truthful when it comes to how you intend to use the engine in question. If you intend to build a daily driver, keep the duration short with an eye toward decent lift for the length of the duration. Don't succumb to the temptation to put the biggest cam you can find into your daily driver. All the cam companies offer comprehensive cam selection procedures based on compression, cruise rpm at 60 mph, and transmission type. This is an attempt to establish a torquey cam selection that will not kill low-speed power with too much overlap.

Conclusion

We really needed about 10 more pages to adequately cover the topic of camshaft basics, but short of that, lift and duration are the most critical points to understanding cam timing. Proper camshaft timing is critical to extracting the most power from any engine and it's difficult to hit the power bull's-eye on the first try. If camshafts seem a bit confusing, you're on the right track to zeroing in on understanding the basics. Once you've mastered the basics, you can then move on to more complex issues like overlap in the next chapter, asymmetrical lobes, inverted flank roller profiles, and intriguing stuff like that. Hey, it's gotta be more fun than collecting butterflies.

Cam Basics

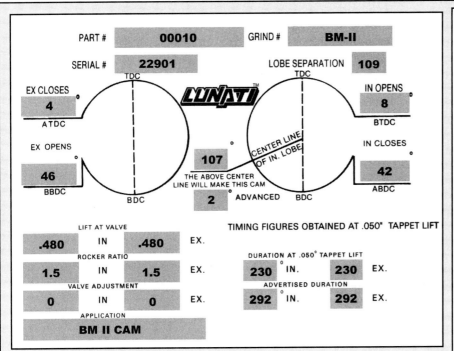

Each manufacturer creates its own cam timing card. This is a Lunati cam card for a single-pattern 230 at 0.050 hydraulic cam. The top half of the card indicates the opening and closing points of the exhaust and intake lobes at the 0.050-inch checking height. Notice that the cam is ground 2 degrees advanced since the intake centerline (A) is at 107 degrees while the lobe separation angle (B) is 109 degrees. The lower half of the card identifies valve lift for the intake and exhaust as well as both advertised and 0.050-inch duration figures. A cam timing card will tell you just about everything you need to know about a cam.

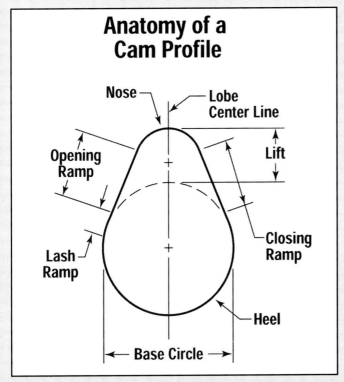

This is a drawing of a cam profile that shows various areas of a typical cam lobe.

Straight Up

If you want to do the job right, you really should degree the cam after it's installed. The task is not difficult, but it does require a couple of hours of work. All you need is a piston stop, degree wheel, pointer, and a crank nut or similarly secure way to turn the engine over.

Some cam companies specify an intake centerline figure, but the easiest way to determine where your cam is installed is to look at the intake opening point at 0.050-inch tappet lift. For example, let's say the cam card calls for the intake to open at 8 degrees BTDC but your measurements indicate the intake opens at 5 degrees BTDC. This means the intake is actually opening later (closer to TDC). To fix this, you will need to advance the cam 3 degrees to open the intake at 8 degrees BTDC. Conversely, let's say the intake opens at 11 degrees BTDC. Then you would need to retard the cam by 3 degrees. If this sounds confusing, take a look at the cam timing graph on page 153.

Advance

Begins intake event sooner
Opens intake valve sooner
More low-end torque
Decreases piston-to-I-valve clearance
Increases piston-to-E-valve clearance

Retard

Delays intake event
Opens intake valve later
More top-end horsepower
Increases piston-to-I-valve clearance
Decreases piston-to-E-valve clearance

NEW VS. OLD

Much of the high performance game is played with an emphasis on nostalgia. Despite the fact the Chevy designed the L79, 350hp 327 cam close to 40 years ago, that cam still enjoys an enthusiastic following. If all you want is a lumpy idle, then this is a great cam. But if you're into high performance for the latest technology that will make more power along with outstanding part-throttle performance, then there are better cams out there.

The following chart compares the L79 cam with a similar Crane PowerMax cam that we chose based on a similar intake valve lift. The big difference is the advertised duration. The L79 cam needs 320 degrees of seat timing to accomplish 0.447-inch lift. The Crane PowerMax 266 needs only 266 degrees to build almost the same lift, a difference of an astounding 54 degrees on the intake side. Also note that even Crane's 0.050-inch intake duration spec is 11 degrees shorter. The Crane would certainly deliver a ton more torque and probably similar peak horsepower numbers and be much more streetable. You can use this same technique when comparing similar cams. Look for the most lift for the least duration, and also look for a cam with the least amount of duration spread between the advertised and 0.050-duration numbers. These cams tend to be the newest designs.

Cam	Advertised Duration	Duration @ 0.050	Lift	Lobe Separation Angle
Chevy L79	320/320	221/221	0.447/0.447	114
Crane PowerMax	266/272	210/216	0.440/0.454	114

Cam Overlap

Understanding Cam Overlap
By Jeff Smith
Photography by Jeff Smith

27

In the early days of high-performance engine building, hot rodders discovered that when they squeezed the intake and the exhaust lobes closer together, the engine responded with more power. That experimentation continues to this day as engine builders struggle with the relationship between overlap and power.

What Exactly Is Overlap?

Before we dive deeply into this subject, we should first define the term. Overlap is the number of crankshaft degrees that both the intake and exhaust valves are open as the cylinder transitions through the end of the exhaust stroke and into the intake stroke. The easiest way to understand overlap is by graphing the two lobes (see graph A, p. 160) and examining the triangle created by the overlap of the intake opening (IO) and the exhaust closing (EC) points. This triangle is measured in degrees based on tappet checking height. This area is also referred to as the lobe separation angle—expressed as the spread in camshaft degrees between the intake centerline and the exhaust centerline.

This area is worth exploring because it is so complex. Let's start with short-duration cams, which offer less overlap than long-duration cams even when they have the same lobe separation angle. Stated another way, a cam with more advertised duration is guaranteed to have more overlap

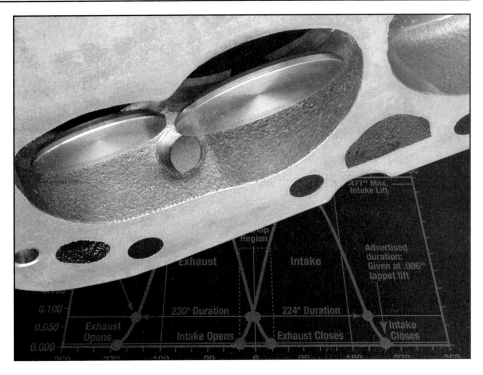

than a cam with less advertised duration even with the same lobe separation angle. It's also important to remember that overlap is ground into the camshaft when it is machined and cannot be changed without grinding a new camshaft unless you have a dual overhead cam engine with separate intake and exhaust cams.

Let's look at a cam with an intake lobe centerline of 106 degrees after top dead center (ATDC) and an exhaust lobe with a 118-degree centerline before top dead center (BTDC). Adding the two values together and dividing by two equals 112 degrees. This is the lobe separation angle. While this number is useful, it doesn't tell us much about the actual overlap between the intake and exhaust lobes. In order to calculate the actual overlap in crankshaft degrees, we need the intake opening (IO) and exhaust closing (EC) points. The SAE spec for advertised duration is 0.006 inch. Keep in mind that all cam companies do not use this spec. Let's take a hydraulic-roller cam with an advertised duration of 276/282 degrees. The IO is 32 degrees BTDC and the EC is 27 degrees ATDC. Add these two values and we come up with 59 degrees of valve overlap.

Now let's take a second, longer-duration cam with advertised duration figures of 294/300 degrees with an IO of 41 degrees BTDC and an EC of 36

Small-Block Chevy Engine Buildups

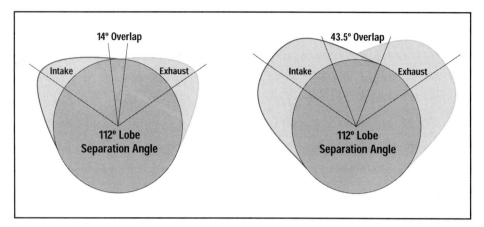

Often an illustration can tell the story better than words. The cam on the left uses relatively short duration lobes with a 112-degree lobe separation angle. The cam on the right uses the same 112-degree lobe separation angle but employs longer-duration intake and exhaust lobes. Note that the cam on the right offers significantly more overlap with 43.5 degrees compared to the smaller cam's 14 degrees.

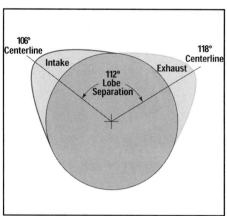

Computing the lobe separation angle is easy. Merely add the intake and exhaust lobe centerlines together and divide by two. In this instance, the intake uses a 106-degree centerline with 118-degree exhaust. The lobe separation angle is 112 degrees (106 + 118 = 224/2 = 112). Unless you are working with a dual overhead cam engine, grinding a new cam is the only way to change the lobe separation angle.

Increasing the rocker ratio does more than just add peak valve lift. Higher-lift rockers also increase overlap by adding lift at all points within the overlap triangle. This has the effect of reducing idle vacuum.

degrees ATDC. We did the math and came up with a much larger figure of 77 degrees even though the lobe separation angle remains at 110 degrees. See how that works? This also illustrates why long-duration camshafts have such a lopey idle with very low idle vacuum. It's not the duration itself, but the number of crankshaft degrees of engine rotation where both the intake and exhaust valves are open at the same time. At idle, there is plenty of time for residual exhaust gas in the combustion chamber to enter the intake manifold when the intake valve opens 41 degrees BTDC. This dilution of the intake manifold with exhaust gas is much like built-in exhaust gas recirculation (EGR) and is the culprit responsible for the unstable idle.

An unfortunate result of excessive overlap is reduced torque and soft throttle response at low engine speeds (e.g., below 3,000 rpm). So why build in all this overlap? The answer can be found when you look into the time that is compressed at high engine speed. At idle, there is plenty of time for the exhaust gas to move back up the intake tract and dilute the incoming charge. The intake air charge is also moving at a very slow speed. But buzz the engine to 6,000 rpm and there is precious little time for the exhaust gas to do anything except exit past the exhaust valve. Add the inertia of high-speed air entering the cylinder when the intake valve opens, and overlap is very useful for initiating that column of air into the cylinder. That 41 degrees BTDC intake opening point now works well to fill the cylinder with a big charge of air and fuel that now makes great power at 6,000 rpm. In

Cam Overlap

Most factory camshafts use wide lobe separation angles to improve idle stability. The '98-2000 LS1 Camaro hydraulic roller cam uses a 122-degree intake centerline combined with a 117-degree exhaust centerline to create a 119.5-degree lobe separation angle. This combines with intake/exhaust duration numbers of 201/212 degrees at 0.050-inch tappet lift. This cam creates very little overlap yet offers excellent midrange torque.

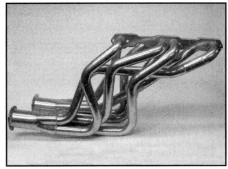

Headers can have a big effect on the power curve since primary pipe size and length help determine when that reflected wave arrives in the cylinder. Power increases when the reflected wave is timed to coincide with camshaft overlap.

effect, that early intake opening point gives the intake system a head start to fill the cylinder at high rpm when there is very little time. The result is more power at these higher engine speeds.

The difficulty with overlap is that the results change with different engine speeds. Since race engines tend to operate in relatively narrow rpm bands (e.g., 5,000 to 7,500 rpm), it's easier to design a cam to work in this rpm band. A street engine is a greater challenge because it must operate through a rpm band of 5,000 rpm or more (1,000 to 6,000 rpm). The key to making overlap work is maximizing the power in the rpm band where you want it. Long overlap periods work best for high-rpm power. For the street, a long overlap period combined with long-duration profiles combine to kill low-speed torque. This makes for a soggy street engine at low engine speeds. Reducing overlap on a long-duration cam will often increase midrange torque at the expense of peak power, but if the average torque improves, that's probably a change worth making.

The most important point in the four-stroke cycle is the intake closing point. While this is not part of overlap, the timing of intake opening and closing determines total duration. The intake closing point is a big determiner in where the engine makes power. A later intake closing point improves top-end power. Combine that with more overlap and you have a cam designed to make power at high rpm. However, it's possible to decrease overlap by using a shorter-duration intake lobe and retard the intake centerline (which spreads the lobe separation angle) to improve midrange power.

We should also look at cams with a short duration and a wide lobe separation angle. All late-model Chevrolet engines use extremely wide lobe separation angles to improve idle quality. A late intake centerline combined with an early exhaust centerline and short-duration lobes creates very little overlap, yet the new LS1 and LS6 Gen III engines make great overall power without having to rely on large overlap periods. This is something to think about.

We would be remiss in not mentioning that many enthusiasts purchase a camshaft strictly on the basis of how it sounds. A cam with generous overlap creates that distinctive choppy idle that just sounds cool. There is a possibility that decreased overlap combined with an idealized intake closing point would create more power while producing a more stable (less lumpy) idle quality. This may not produce the idle sound most enthusiasts want to hear, but it is intriguing nonetheless.

There's a ton more to learn about overlap and lobe separation angles than we can really get into in this short amount of space. It's a relatively complex subject with many different conflicting requirements. But the more you learn about camshafts and how they operate, the more power you can make from your street engine.

Small-Block Chevy Engine Buildups

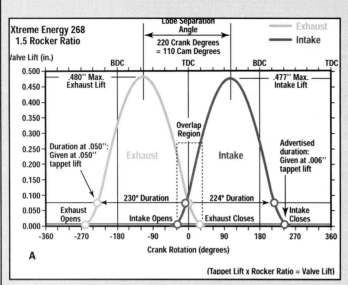

A

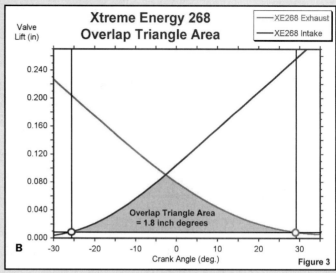

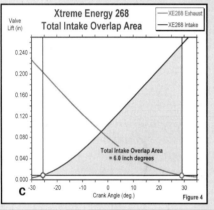

C — Figure 4

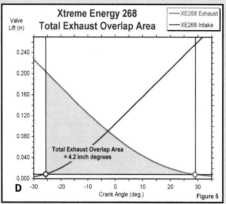

D — Figure 5

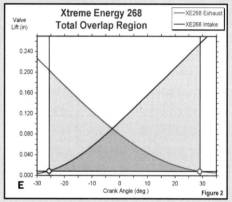

E — Figure 2

Besides the number of degrees of overlap, we can also graph overlap area. These graphs, courtesy of Comp Cams lobe builder Billy Godbolt, take a closer look at the no-man's-land of overlap. Let's look at the first of the four enlarged overlap areas of a Comp Cams Xtreme Energy 268 flat-tappet hydraulic cam shown in graph B.

The overlap triangle in the center is what you usually see in overlap graphs. This graph represents the overlap area bordered by the exhaust closing and intake opening ramps. This area represents the most active area of the overlap period where both valves are open around TDC.

In graph C, note that the shaded area deals with the total intake overlap that begins the moment the intake valve comes off its seat until the exhaust valve closes. At the instant the intake valve opens, the exhaust valve is around 0.235 inch off the seat and closing. As the crankshaft turns toward TDC, the intake continues to open, exposing more of the intake tract to the combustion chamber while the exhaust valve is still open. The two lift curves cross over at around 0.080-inch valve lift, roughly 4 degrees BTDC.

This crossover point can be moved several different ways. You can move the intake centerline toward TDC, which will increase overlap, or move it away, which decreases the overlap area. Or you could move the exhaust-lobe centerline in or out. Moving the intake centerline probably has the most effect since it will move the intake closing point.

This overlap area is important; if the exhaust port or exhaust system is restrictive, it's more than likely that there will be residual exhaust pressure in the combustion chamber that will try to push up into the intake port and into the intake manifold. Since exhaust gas will not combust a second time, this residual exhaust gas will dilute the incoming air/fuel charge and reduce cylinder pressure. However, if the exhaust port and system are efficient, the exiting exhaust gas will create an additional signal (with a properly timed reflected wave—see "Scavenging" sidebar) to help pull more air and fuel into the cylinder.

The shaded portion of graph D illustrates the total exhaust overlap area where the exhaust system is opened to the chamber with the intake valve open. Note that the shaded areas of the intake and exhaust overlap areas are not the same. This is because of the different lobe profiles used for the intake and exhaust lobes.

Graph E incorporates all three shaded areas. For example, at 20 degrees ATDC, the exhaust valve is still approximately 0.020 inch off the seat while the intake valve is open 0.210 inch. The risk with overlap is that it's possible that even at this point, where the exhaust valve is almost closed, that some portion of the incoming intake charge could escape out the exhaust. If the exhaust valve is closed sooner, this will affect the engine throughout the entire rpm curve. An earlier closing exhaust valve would reduce overlap and help low-speed torque, while reducing top-end power. That's why each engine combination requires individual testing to determine the best overall power curve.

Scavenging

In order to understand what happens in the combustion space during overlap, we need to look at an interesting phenomenon we'll call wave tuning. If you toss a rock into a pond, the impact will cause waves to radiate outward from where the rock struck the water. When the exhaust valve opens after combustion, a similar pressure wave is created and exhaust gas flows out the header tube until it exits an open header. As this positive pressure wave exits the header collector, this creates a negative pressure wave that is reflected back up the header pipe toward the cylinder. This negative pressure wave actually improves exhaust gas flow that is still occurring. The time it takes this negative-pressure wave to arrive in the chamber (the exhaust valve is still open) is determined by the length of the header tube and engine rpm.

If the negative-pressure wave arrives in the cylinder during overlap, it adds additional negative pressure (vacuum) to the cylinder, allowing atmospheric pressure above the carburetor to push more air and fuel into the cylinder. The timing and strength of this reflected wave action is determined by several factors including header length, primary header tube diameter, header collector size and length, cam overlap, and at least a couple dozen more details too mind-numbing to mention.

It's difficult to assign a priority to any one or two items that contribute to a power increase with relation to overlap. This is a much more complex subject than our simplistic explanation might suggest. While this short explanation will probably create more questions than it answers, we felt it was important to mention the existence of reflected waves to help explain why overlap contributes to the power curve. If you are interested in learning more about wave tuning, there is a computer simulation program called Dynomation from VP Engineering that incorporates this phenomenon and gives a full explanation of wave dynamics in its accompanying material. Contact VP Engineering for more information.

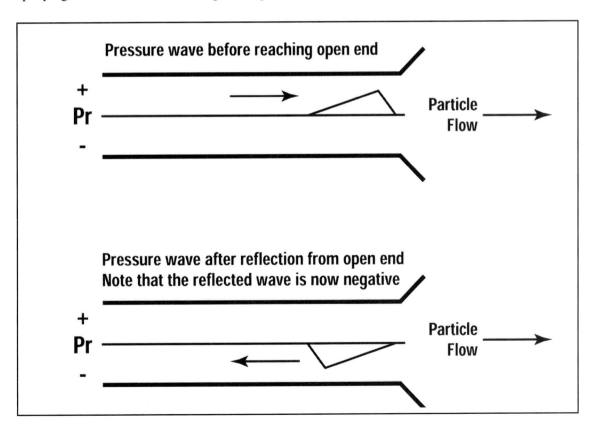

28 Roller Cam Basics

Fact vs. Fiction On Roller Cams
By Jeff Smith
Photography by Jeff Smith

The new darling of the camshaft world is the hydraulic-roller cam. GM made it fashionable back in the late '80s when Chevy put hydraulic roller cams in production Corvettes and '87 Camaro engines. At first, the performance community viewed this innovation as a low performance option, but cam companies soon released performance versions of these cams that are now making serious horsepower. We decided to take a look at what makes roller cams so powerful and popular.

Before getting into the specifics, let's take a look at some basic cam tech to find out why a roller works so well. A hydraulic roller cam offers much more aggressive lift-curve capabilities compared to a flat-tappet cam. In order for a flat-tappet cam to generate as much lift as a roller, it requires more duration. Flat-tappet lifters will actually dig into the lobe flank if the lobe is designed too aggressively. Roller tappets do not suffer that problem, so the designer can put much more lift into a roller-cam lobe. The accompanying "Profiling" graph makes this easier to understand.

Curves Ahead

If you compare a hydraulic-roller cam to a flat-tappet hydraulic cam with similar duration at 0.050-inch numbers, the hydraulic-roller cam will always have a longer seat-duration figure. While the roller configuration allows faster ramp acceleration, it also suffers from slow acceleration off the seat compared to a flat-tappet cam. Therefore, the advertised-duration numbers are slightly longer.

This means that a hydraulic-roller cam with the same duration at 0.050-inch lift as a flat-tappet hydraulic will not idle exactly the same. If you read the last chapter story on overlap, then you know that more advertised duration increases the amount of overlap, which will cause a somewhat lumpier idle. This is not a big problem, but worth noting if you are considering swapping in a hydraulic-roller cam. If a smoother idle is important, the roller cam could easily be ground with a wider lobe separation angle (110 to 114 degrees for example) to reduce the overlap.

Roller Cam Basics

It's obvious from the shape of a hydraulic-roller cam lobe compared to a flat-tappet that the roller offers more duration at the higher lift. The flat-tappet cam lobe is more pointed while the roller cam nose is more blunt, which means the roller is holding the valve open at higher lift longer. That's a good thing.

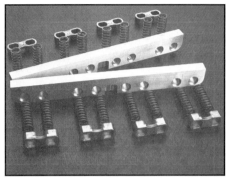

The biggest problem with hydraulic-roller cams is that the tall lifters are heavy. The added weight makes it difficult to control at high rpm without using stiffer valvesprings. Unfortunately, heavy springs tend to collapse hydraulic lifters at high rpm. AirFlow Research sells a Hydra-Rev kit for the small-block Chevy that adds rpm by adding spring pressure to control the lifter body.

Because a roller lifter can accommodate a more aggressive lift curve, a roller cam can produce much more lift than a flat-tappet cam for the same duration.

Powerful Profiles

At first it might appear that the real key to a hydraulic-roller cam is the additional lift. This is true, as we point out in the "Profiling" sidebar. Adding this additional duration above 0.200-inch tappet lift is really aimed at increasing airflow. The true advantage of a hydraulic or mechanical roller is that the profile can hold the valve open longer during the time that the cylinder head offers the most potential flow.

While the right hydraulic-roller cam will improve power even on an engine with stock heads, the real potential lies in combining a roller cam with an engine equipped with a set of good-flowing cylinder heads. While this includes monsters with huge ports, you should not overlook even mild heads like the iron Vortecs. These heads offer outstanding mid-lift flow numbers in the 0.200- to 0.400-inch valve-lift range. Slide in a roller cam with longer 0.200-inch duration and avoid long advertised duration and a late intake closing and you have a recipe for incredible power in an engine that is still very streetable.

Consider that the Vortec head does not offer killer flow numbers above 0.500-inch lift. What it offers instead is outstanding flow between 0.200- and 0.500-inch valve lift. This is where the hydraulic-roller cam shines. Combine the two and you have a powerful combination. Of course, there are other heads that also offer this same kind of great mid-range flow potential, but you can buy a complete pair of Vortec heads for under $450. Use the money you save with the heads to purchase a hydraulic roller cam package and you're on your way to big-time power.

Ups and Downs

There are some great reasons for stepping up to a hydraulic-roller cam package, but all is not perfect in the roller-cam world. One glance at the pricing sidebar will reveal that this better technology comes at a hefty price. This price includes adding in

163

Profiling

This chart compares the timing curves of a hydraulic flat-tappet cam to a hydraulic-roller cam with matching duration figures at 0.050. Notice that the roller requires a longer seat duration (advertised duration) with the same duration at 0.050-inch tappet lift. We've also added a third measurement listing duration at 0.200-inch tappet lift that you may not have seen before. Note that the hydraulic roller offers 145 degrees of duration versus the flat-tappet's 137 degrees.

The horsepower key is to combine the roller cam's more aggressive lift curve with a good cylinder head that flows well between 0.200- and 0.500-inch valve lifts. This is the real reason why roller cams make more power. We also looked at specs on a mechanical-roller cam. A Comp solid roller at 224 degrees of duration at 0.050 delivers 147 degrees of duration at 0.200-inch tappet lift, plus another 0.038-inch of lift over the hydraulic-roller cam with 0.540-inch lift with a 1.5:1 rocker. Yahoo.

Cam	Advertised Duration	Duration @ 0.050	Int. Dur. @ 0.200	Lift
XE268 Hyd. flat-tappet	268/280	224/230	137	0.477/0.480
XE276 Hyd. roller	276/282	224/230	145	0.502/0.510

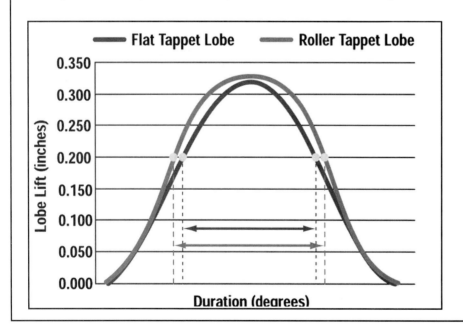

more expensive roller rocker arms, stronger pushrods, and better valve springs. If you plan to bolt a hydraulic roller cam into a pre-'87, non-roller cam block, the retrofit lifter kits can get expensive. This will also require a thrust bearing on the front of the cam to prevent cam walk.

This isn't required with a flat-tappet cam since the lobes are cut at a slight angle to offset the cam thrust movement that's inherent in a roller cam. This is just one more piece to the hydraulic roller cam investment portfolio that's important to know.

One way to control this additional expense on your next engine buildup is to start with an '87-or-later hydraulic roller cam block. For example, you can pick up one of these inexpensive short-blocks as the starting point. This allows you to use the factory hydraulic lifters and retaining system, which can be less expensive than the aftermarket retrofit kits.

Bottom Line

As usual, if you want to make more power, it's gonna cost more money. The good news is that a properly selected hydraulic roller cam offers the potential to make more horsepower and more torque without sacrificing much in the way of street manners. Assemble the right combination of parts and you might just find yourself suffering from massive traction problems. That's a good problem to have.

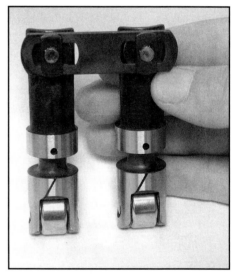

Mechanical roller cams offer even more lift and area under the curve than hydraulic rollers, but can suffer from durability problems if used in high-mileage applications. Comp's Endure-X cure is to aim lube directly at the lifter roller. Excessive spring pressures are still the ultimate problem for big-lift cams on the street.

Roller Cam Basics

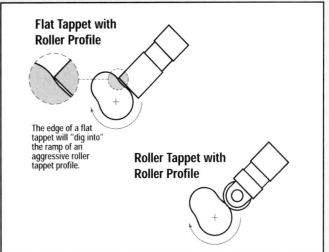

Roller cams require a much harder steel surface than flat-tappet cams because of higher loads. In the early days of roller cams, this meant you had to run a soft silicon-bronze gear that wore out quickly. Now, most cam companies offer iron gears on the end of the steel shaft to accommodate a standard iron distributor gear.

The edge of a flat tappet will "dig into" the ramp of an aggressive roller tappet profile.

This illustration shows how a flat-tappet lifter will dig into an aggressive roller profile, which is why flat-tappet cams cannot offer the same profile as a roller cam.

Hydraulic Roller Cams

The following are performance hydraulic-roller cams. All cam lift numbers are calculated with a 1.5:1 rocker ratio. A 1.6 ratio would add approximately 0.030 inch of valve lift.

Cam	Adv. Dur. (degrees)	Dur. @ 0.050 (degrees)	Lift (inches)	Lobe Sep. (degrees)
Comp Cams				
XR264HR	264/270	212/218	0.478/0.495	110
XR270HR	270/276	218/224	0.495/0.502	110
XR276HR	276/282	224/230	0.502/0.510	110
XR282HR	282/288	230/236	0.510/0.520	110
XR288HR	288/294	236/242	0.520/0.540	110
XR294HR	294/300	242/248	0.540/0.562	110
Crane Cams				
Compu 2032	270/276	214/220	0.452/0.465	112
PowerMax 284	284/292	222/230	0.509/0.528	112
PowerMax 296	296/304	234/242	0.539/0.558	112
PowerMax 302	302/306	240/244	0.558/0.558	110
GM Performance Parts				
24502586 (HOT)	279/287	218/228	0.492/0.492	112
12370846	NA/NA	222/230	0.509/0.528	112
12370847	NA/NA	234/242	0.539/0.558	112
Isky				
RR272/282	272/282	225/234	0.505/0.530	112
RR282/294	282/294	234/246	0.530/0.550	110
Lunati				
54755	268/276	215/218	0.489/0.503	115
54779LUN	268/279	215/224	0.489/0.489	112
54743LUN	287/298	219/227	0.471/0.480	112
54761LUN	298/286	227/234	0.478/0.480	112
54756LUN	290/300	232/242	0.510/0.525	110

Source Index

AE Clevite
1350 Eisenhower Pl.
Ann Arbor, MI 48108-3282
734/975-4777
www.engineparts.com

Air Flow Research (AFR)
10490 Ilex Ave.
Pacoima, CA 91331-3137
818/890-0616
www.airflowresearch.com

Air Lift Co.
P.O. Box 80167
Lansing, MI 48908-0617
800/248-0892
www.airliftcompany.com

ATI Performance Products
6747 Whitestone Rd.
Baltimore, MD 21207
410/298-4343

Automotive Racing Products (ARP)
531 Spectrum Cir.
Oxnard, CA 93030
800/826-3045
805/278-RACE (7223)

Autotronic Controls Corp. (MSD)
1490 Henry Brennan Dr.
El Paso, TX 79936
915/857-5200
www.msdignition.com

Barry Grant Fuel Systems
1450 McDonald Rd.
Dahlonega, GA 30533
706/864-4712
www.barrygrant.com

Brodix
301 Maple St.
Mena, AR 71953
501/394-1075
www.brodix.com

Canfield Heads
580 W. Main St.
Canfield, OH 44406
330/533-7092
www.canfieldheads.com

Canton Racing Products
2 Commerce Dr.
North Branford, CT 06471-1250
203/484-4900

Childs & Albert
24849 Anza Dr.
Valencia, CA 91355
661/295-1900
www.childs-albert.com

Coast High Performance
1650 W. 228th St.
Torrance, CA 90501
310/784-2977
www.coasthigh.com

Cola Cranks
19122 S. Santa Fe
Rancho Dominquez, CA 90221
310/639-7700

Competition Products
(source for Canfield heads)
280 W. 35th Ave.
Oshkosh, WI 54902
920/233-2023
800/233-0199 (orders)

Comp Cams
3406 Democrat Rd.
Memphis, TN 38118
901/795-2400
800/999-0853 Cam Help

CSI Performance Products
16936 County Rd. 252
McAlpin, FL 32062
904/776-1476
800/226-1274

Dart Machinery
353 Oliver Dr.
Troy, MI 48084
248/362-1188
www.dartheads.com

Demon Carburetion
1450 McDonald Rd.
Dahlonega, GA 30533
706/864-4712
www.demoncarbs.com

Edelbrock
2700 California St.
Torrance, CA 90503
310/781-2222
310/782-2900 Tech line
www.edelbrock.com

Federal-Mogul Corp. (Carter, Fel-Pro, Speed-Pro)
26555 Northwestern Hwy.
Southfield, MI 48034
248/354-7700
www.federal-mogul.com

Gellner Engineering
2827 Brookpark Rd.
Parma, OH 44134
216/398-8500

GM Performance Products
800/577-6888 for your nearest GM Performance Parts dealer

Hedman Hedders
16410 Manning Way
Cerritos, CA 90703

Source Index

562/921-0404
www.hedman.com

Holley Performance Products
P.O. Box 10360
Bowling Green, KY 42102-7360
502/782-2900
800/465-5391 Nearest dealer
502/781-9741 Tech

Hooker Headers
1024 W. Brooks
Ontario, CA 91762
909/983-5871

Hye Tech Performance
3475 N. Peck Rd.
El Monte, CA 91731
626/575-0053

Iskenderian Racing Cams (Isky)
16020 S. Broadway St.
Gardena, CA 90248
323/770-0930
www.Iskycams.com

JE Pistons
15312 Connector Ln.
Huntington Beach, CA 92649
714/898-9763
www.jepistons.com

Jim Grubbs Motorsports
28130 Ave. Crocker, Unit 331
Valencia, CA 91355-4629
661/257-0101

Ken Duttweiler Performance
1563 Los Angeles Ave.
Saticoy, CA 93004
805/659-3648

Made For You Products
P.O. Box 720700
Pinon Hills, CA 92372
760/868-6962

Manley Performance Products
1960 Swarthmore Ave.
Lakewood, NJ 08701
732/905-3366
www.manleyperformance.com

McKenzie's Cylinder Heads
534 Montgomery Ave., Unit 206
Oxnard, CA 93030
805/485-1810

Melling Automotive Products
2620 Saradan Dr.
Jackson, MI 49204
517/787-8172
www.melling.com

Mickey Thompson Performance Tires
4670 Allen Rd.
Stow, OH 44224
800/222-9092
330/928-9092 (tech line)
www.mickeythompsontires.com

Milodon Inc.
20716 Plummer St.
Chatsworth, CA 91311
818/407-1211
www.milodon.net

Moroso Performance Products
80 Carter Dr.
Guilford, CT 06437-0570
203/453-6571
www.moroso.com

Mr. Gasket (Mallory)
10601 Memphis Ave., Ste. 12
Cleveland, OH 44144
216/688-8300
888/MRGASKET ext. 9999
www.mrgasket.com

Performance Automotive Warehouse
21001 Nordhoff St.
Chatsworth, CA 91311
818/678-3000 Orders and Tech

Plasti-Kote
1000 Lake Rd.
Medina, OH 44258
800/431-5928

Probe Industries
1650 W. 228th St.
Torrance, CA 90501
310/784-2977

Proform
26708 Grossbeck Hwy.
Warren, MI 48089
800/521-1005
810/774-2500
www.proformparts.com

Scat Enterprises
1400 Kingsdale Ave.
Redondo Beach, CA 90278
310/370-5501
www.scatenterprises.com

Scoggin-Dickey Parts Center
5901 Spur 327
Lubbock, TX 79424-2705
800/456-0211
806/798-4108 (tech)
www.scoggindickey.com

SI Valves
2175-A Agate Ct.,
Simi Valley, CA 93065
800/564-8258
805/582-0085

Standard Abrasives
4201 Guardian St.
Simi Valley, CA 93063
800/383-6001

Southside Machine Co.
6400 N. Honeytown Rd.
Smithville, OH 44677
330/669-3556

Small-Block Chevy Engine Buildups

Sportsman Racing Products
15312 Connector Ln.
Huntington Beach, CA 92649
714/898-9763
www.jepistons.com

Summit Racing Equipment
P.O. Box 909
Akron, OH 44309-0909
800/230-3030 Orders
330/630-0230 Customer service
www.summitracing.com

Tim Moore Automotive
10953 Tuxford, Unit 6
Sun Valley, CA 91352
818/249-2363

Trick Flow Specialties (TFS)
1248 Southeast Ave.
Tallmadge, OH 44278
330/630-1555

Ventura Motorsports
(Ed Taylor)
P.O. Box 33
Ventura, CA 93002
805/485-3609 Shop

Vibratech Inc. (Fluidampr)
11980 Walden Ave.
Alden, NY 14004
716/937-7903
www.vibratech.com

Westech Performance Group
11098 Venture Dr., Ste. C
Mira Loma, CA 91752
909/685-4767

World Products
35330 Stanley
Sterling Heights, MI 48312
810/939-9628
www.worldcastings.com

Zoops Products
931 E. Lincoln St.
Banning, CA 92220
909/922-2396
www.zoops.com

OTHER BOOKS BY HPBOOKS

HANDBOOKS
Auto Electrical Handbook: 0-89586-238-7 or HP1238
Auto Upholstery & Interiors: 1-55788-265-7 or HP1265
Car Builder's Handbook: 1-55788-278-9 or HP1278
The Lowrider's Handbook: 1-55788-383-1 or HP1383
Powerglide Transmission Handbook: 1-55788-355-6 or HP1355
Turbo Hydramatic 350 Handbook: 0-89586-051-1 or HP1051
Welder's Handbook: 1-55788-264-9 or HP1264

BODYWORK & PAINTING
Automotive Detailing: 1-55788-288-6 or HP1288
Automotive Paint Handbook: 1-55788-291-6 or HP1291
Fiberglass & Composite Materials: 1-55788-239-8 or HP1239
Metal Fabricator's Handbook: 0-89586-870-9 or HP1870
Paint & Body Handbook: 1-55788-082-4 or HP1082
Pro Paint & Body: 1-55788-394-7
Sheet Metal Handbook: 0-89586-757-5 or HP1757

INDUCTION
Bosch Fuel Injection Systems: 1-55788-365-3 or HP1365
Holley 4150: 0-89586-047-3 or HP1047
Holley Carbs, Manifolds & F.I.: 1-55788-052-2 or HP1052
Rochester Carburetors: 0-89586-301-4 or HP1301
Turbochargers: 0-89586-135-6 or HP1135
Weber Carburetors: 0-89586-377-4 or HP1377

PERFORMANCE
Baja Bugs & Buggies: 0-89586-186-0 or HP1186
Big-Block Chevy Performance: 1-55788-216-9 or HP1216
Big-Block Mopar Performance: 1-55788-302-5 or HP1302
Bracket Racing: 1-55788-266-5 or HP1266
Brake Systems: 1-55788-281-9 or HP1281
Camaro Performance: 1-55788-057-3 or HP1057
Chassis Engineering: 1-55788-055-7 or HP1055
Chevy Trucks: 1-55788-340-8 or HP1340
Ford Windsor Small-Block Performance: 1-55788-323-8 or HP1323
4-Wheel & Off-Road's Chassis/Suspension: 1-55788-406-4/HP1406
Honda/Acura Engine Performance: 1-55788-384-X or HP1384
High Performance Hardware: 1-55788-304-1 or HP1304
How to Hot Rod Big-Block Chevys: 0-912656-04-2 or HP104
How to Hot Rod Small-Block Chevys: 0-912656-06-9 or HP106
How to Hot Rod Small-Block Mopar Engines: 0-89586-479-7 or HP1479
How to Hot Rod VW Engines: 0-912656-03-4 or HP103
How to Make Your Car Handle: 0-912656-46-8 or HP146
John Lingenfelter: Modify Small-Block Chevy: 1-55788-238-X or HP1238
Mustang 5.0 Projects: 1-55788-275-4 or HP1275
Mustang Performance (Engines): 1-55788-193-6 or HP1193
Mustang Performance 2 (Chassis): 1-55788-202-9 or HP1202
Mustang Perf. Chassis, Suspension, Driveline Tuning: 1-55788-387-4 or HP1387
Mustang Performance Engine Tuning: 1-55788-387-4 or HP1387
1001 High Performance Tech Tips: 1-55788-199-5 or HP1199
Performance Ignition Systems: 1-55788-306-8 or HP1306
Small-Block Chevy Performance: 1-55788-253-3 or HP1253
Small Block Chevy Engine Buildups: 1-55788-400-5 or HP1400
LS1/LS6 Small-Block Chevy Performance: 1-55788-407-2 or HP1407

ENGINE REBUILDING
Engine Builder's Handbook: 1-55788-245-2 or HP1245
How to Rebuild Small-Block Chevy LT-1/LT-4: 1-55788-393-9/HP1393
Rebuild Air-Cooled VW Engines: 0-89586-225-5 or HP1225
Rebuild Big-Block Chevy Engines: 0-89586-175-5 or HP1175
Rebuild Big-Block Ford Engines: 0-89586-070-8 or HP1070
Rebuild Big-Block Mopar Engines: 1-55788-190-1 or HP1190
Rebuild Ford V-8 Engines: 0-89586-036-8 or HP1036
Rebuild GenV/Gen VI Big-Block Chevy: 1-55788-357-2 or HP1357
Rebuild Small-Block Chevy Engines: 1-55788-029-8 or HP1029
Rebuild Small-Block Ford Engines: 0-912656-89-1 or HP189
Rebuild Small-Block Mopar Engines: 0-89586-128-3 or HP1128

RESTORATION, MAINTENANCE, REPAIR
Camaro Owner's Handbook ('67–'81): 1-55788-301-7 or HP1301
Camaro Restoration Handbook ('67–'81): 0-89586-375-8 or HP1375
Classic Car Restorer's Handbook: 1-55788-194-4 or HP1194
How to Maintain & Repair Your Jeep: 1-55788-371-8 or HP1371
Mustang Restoration Handbook ('64 1/2–'70): 0-89586-402-9 or HP1402
Tri-Five Chevy Owner's Handbook ('55–'57): 1-55788-285-1 or HP1285

GENERAL REFERENCE
A Fan's Guide to Circle Track Racing: 1-55788-351-3 or HP1351
Auto Math Handbook: 1-55788-020-4 or HP1020
Corvette Tech Q&A: 1-55788-376-9 or HP1376
Ford Total Performance, 1962–1970: 1-55788-327-0 or HP1327
Guide to GM Muscle Cars: 1-55788-003-4 or HP1003

MARINE
Big-Block Chevy Marine Performance: 1-55788-297-5 or HP1297
Small-Block Chevy Marine Performance: 1-55788-317-3 or HP1317

ORDER YOUR COPY TODAY!
All books can be purchased at your favorite retail or online bookstore (use ISBN number), or auto parts store (Use HP part number). You can also order direct from HPBooks by calling toll-free at 800/788-6262, ext. 1.